Roadmap to the Virginia SOL
EOC Chemistry

Roadmap
to the Virginia SOL
EOC Chemistry

by
Paul Foglino

Random House, Inc.
New York

www.randomhouse.com/princetonreview

This workbook was written by The Princeton Review, one of the nation's leaders in test preparation. The Princeton Review helps millions of students every year prepare for standardized assessments of all kinds. The Princeton Review offers the best way to help students excel on standardized tests.

The Princeton Review is not affiliated with Princeton University or Educational Testing Service.

Princeton Review Publishing, L.L.C.
160 Varick Street, 12th Floor
New York, NY 10013

E-mail: textbook@review.com

Copyright © 2005 by Princeton Review Publishing, L.L.C.

All rights reserved under International and Pan-American Copyright Convention.

Published in the United States by Random House, Inc., New York.

ISBN 0-375-76442-9

Content Editor: Linda Fan
Series Editor: Russell Kahn
Design Director: Tina McMaster
Development Editor: Scott Bridi
Art Director: Neil McMahon
Production Coordinator: Alexandra Morrill
Production Editor: Evangelos Vasilakis

Manufactured in the United States of America

9 8 7 6 5 4 3 2 1

First Edition

CONTENTS

Introduction 1

Lessons and Reviews

Lesson 1: Test Structure and Strategies 7

Lesson 2: Atomic Structure 15

Review for Lesson 2 25

Lesson 3: The Periodic Table 31

Review for Lesson 3 39

Lesson 4: Chemical Compounds 47

Review for Lesson 4 57

Lesson 5: Chemical Reactions 67

Review for Lesson 5 77

Lesson 6: Chemistry Computation 81

Review for Lesson 6 93

Lesson 7: Gases 107

Review for Lesson 7 113

Lesson 8: Phase Changes 119

Review for Lesson 8 129

Lesson 9: Solutions 135

Review for Lesson 9 143

Lesson 10: Equilibrium 149

Review for Lesson 10 155

Lesson 11: Acids and Bases 163

Review for Lesson 11 171

Lesson 12: Causes of Chemical Reactions 179

Review for Lesson 12 185

Lesson 13: Scientific Investigation 191

Review for Lesson 13 195

The Practice Tests

Instructions for Taking the Practice Tests 203

Practice Test 1 205

Practice Test 1 Answer Sheet 207

Answers and Explanations for Practice Test 1 221

Practice Test 2 231

Practice Test 2 Answer Sheet 233

Answers and Explanations for Practice Test 2 247

INTRODUCTION

THE END OF COURSE EXAMS

The end of what? Well, it's not *the* end. The End of Course (EOC) exams are only the final exams for certain "core" courses offered at your school. The Virginia Department of Education (VDOE) has decided that there are certain skills that it wants you to have when you graduate high school. Therefore, to receive a diploma, you must pass six EOC exams. EOC exams are given in the core subjects: English, math, science, and history and social science. These exams evaluate not only what you have learned, but also how well your school has taught its students.

WHAT EXACTLY IS AN SOL?

You've probably heard the EOC exams referred to as the "SOL tests." SOL stands for Standards of Learning, which is simply the name for the specific set of skills that the VDOE has earmarked for each core subject.

For example, in chemistry, the Standards of Learning state that a student should be able to balance chemical equations. So, rather than just taking your chemistry teacher's word for it, the VDOE has drawn up its own Chemistry exam to make sure you can, in fact, balance chemical equations (among other skills).

> **TIP:** If you want to know more about the VDOE standards, try its Web site www.pen.k12.va.us/VDOE.

If you're feeling a little nervous, don't sweat it. This book is going to ensure that you have mastered every skill that the VDOE expects you to know for the EOC Chemistry test. It will give you practice questions and explain the ins and outs of the test, and it will explain how many questions you need to get right and how many SOL exams you need to pass in order to receive your diploma. In short, you've purchased the only guide to the EOC Chemistry exam you'll ever need. Congratulations for thinking ahead!

WHO IS THE PRINCETON REVIEW?

The Princeton Review is one of the world's leaders in test preparation. We've been preparing students for standardized tests since 1981 and have helped millions reach their academic and testing goals. Through our courses, books, and online services, we offer strategy and advice on the SAT, PSAT, SAT IIs, and the TerraNova, just to name a few. The Princeton Review has created nearly twenty books to help Virginia students with their SOL exams.

WE HAVE THE INSIDE SCOOP

The EOC Chemistry exam is not immune to strategy and preparation. This book includes all of the information about the exam you'll need to do well. The Princeton Review has been looking at standardized tests like the EOC Chemistry exam for years, and we'll share our special techniques to approaching standardized tests and chemistry questions so you'll have every possible opportunity to score your best on this exam.

TIP: We have made sure that every skill listed by the Standards of Learning is reviewed and practiced in this book.

One of the biggest obstacles for students in standardized testing is test anxiety. Taking the EOC exams with your diploma on the line can create stressful conditions. In order to reduce stress and prepare you for this exam, we've dedicated countless hours of research to help you do well on the EOC Chemistry exam. In addition, we've written two practice tests to help you realistically evaluate your skills.

HOW IS THIS BOOK ORGANIZED?

This book has two primary purposes. First, we will familiarize you with the structure of the exam and recommend the soundest test-taking strategies to maximize your score.

Second, we want to make sure that you're familiar with the raw material of the exam—the actual chemistry skills and concepts that must be mastered to do well on the exam. We will focus on exactly how the Virginia EOC exam tests your chemistry skills so you will know what to expect on exam day.

Lesson 1 includes test-taking strategies and techniques that are useful for the EOC Chemistry exam and any other chemistry test you take. Lessons 2–13 review all of the chemistry skills and concepts listed in the Virginia Standards of Learning for Chemistry. The concepts in these lessons are the same ones that are tested by the EOC Chemistry exam.

The book also includes two complete practice tests with answer keys and explanations so that you can assess your skills under exam conditions. After you've worked through this book, there will be no surprises when you take the actual EOC Chemistry test.

WHAT DO THE SOL EXAMS MEAN FOR GRADUATION?

This is a big question, and you should be clear on this before you proceed to the finer points of how to ace this test.

Virginia high school students must pass six of twelve possible EOC exams in order to graduate and receive a Standard Diploma. But if you aren't satisfied with a standard diploma, there's also an Advanced Studies Diploma, which requires that you pass nine of the twelve EOC exams.

Students in Virginia must pass at least one EOC exam in science in order to earn a diploma. Virginia SOL EOC exams in science are Earth Science, Biology, and Chemistry.

If you want more information about an Advanced Studies Diploma, speak to a guidance counselor, teacher, or adminstrator at your school.

FREQUENTLY ASKED QUESTIONS

- **What Does the Test Look Like?**
  The SOL Chemistry exam tests five major skill sets called "Reporting Categories." Each evaluates specific chemistry concepts and skills. Later in the book, you will review all the specific skills you'll need to master to succeed on the test. If you want to peek at them now, you can check out the VDOE Web site at www.pen.k12.va.us.

There are sixty multiple-choice questions on the SOL End-of-Course Chemistry exam. The questions break down according to Reporting Categories in the following way:

Reporting Categories	Number of Questions
Scientific Investigation	10
Atomic Structure and Periodic Relationships	8
Nomenclature, Chemical Formulas, and Reactions	16
Molar Relationships	8
Phases of Matter and Kinetic Molecular Theory	8
Total Number of Scored Questions:	50
Field Test Items:	10
Total Number of Questions:	60

- **What Is a Field Test Item?**
  A field test item is an experimental question that does not count toward your final score. Because you have no way of knowing which questions are field test items and which ones count toward your score, answer every question as if it counts toward your score. The test writers use students' performance on field test items to determine whether they've written good questions for use on future exams.

- **What Is a Passing Score?**
  You have to answer correctly twenty-seven out of the fifty scored questions to pass the EOC Chemistry exam. There are sixty total questions on the exam, but ten of them are field test items that do not count toward your score. These ten questions are mixed together with the fifty scored questions, so you won't be able to tell which questions really count as you take the test. Do yourself a favor and answer **all** the questions as if they count toward your final score.

LESSONS AND REVIEWS

LESSON 1
TEST STRUCTURE AND STRATEGIES

LEARN THE BASICS
There's no getting around it. The only way to get a good score on the Chemistry SOL exam is to learn chemistry. But take heart, this isn't the medical school entrance exam. It's not the AP Chemistry test. It's not the SAT II. The SOL Chemistry exam is easier than any of those tests. In fact, the questions on this test may turn out to be easier than the questions you saw on any of the exams you took in class this year.

The point is that you don't need to master every little detail of chemistry to do well on this test. What you need to do is make sure that you know the basics. That's what this book will help you to do.

This book will help you with other things, as well. After all, this is a standardized test, and standardized tests are used to measure the abilities of students who go to different schools with different teachers with the same test. The most important thing is to realize that taking a standardized test is like playing a game; in order to do well, you need to learn the rules. Here are a few things that can help in improving your scores.

COUNT YOUR BLESSINGS
Rules are normally things that tell you what you can't do when you play a game. In this game, however, you'll find that some of the rules are there to make things easier on you. Here are some reasons why the Virginia SOL is even easier than the average standardized test.

IT'S A MULTIPLE-CHOICE TEST
Multiple choice is a test-taker's best friend. Every question comes with four possible answer choices. In this exam the choices will be labeled **A**, **B**, **C**, and **D** for odd numbered questions and **F**, **G**, **H**, and **J** for even questions. Your job, of course, is to find the correct answer. But don't fret. The very nature of a multiple-choice test lends itself to be an easier exam for students. In fact, a multiple-choice test is a beautiful thing for a couple of reasons.

- What if you know the answer, but it's right on the tip of your tongue and you can't quite remember it. No problem. One quick glance at the answer choices and the correct answer will jump right out at you. You couldn't do that on a fill-in-the-blank test.

- What if you don't know the answer. You can always guess. There are only four possible answers. One of them is right. On a multiple-choice test with four answers, most people can get 25 percent of the questions right even if they don't know a single answer. Those are "free" points. You can get even more free points if you use POE, Process of Elimination, which will be discussed later in this chapter.

It may sound a little silly, but guessing is the most useful and powerful technique for doing well on the SOL Chemistry test. There's one simple reason for that.

There's No Guessing Penalty

Most standardized test makers don't like the idea that you can get free points by guessing randomly. So on many standardized tests, like the SAT or the SAT II, you are penalized for your wrong answers. That is, not only do you not get credit for an incorrect answer, but points are actually deducted when you get one wrong. On a test with a guessing penalty, sometimes it's best to just leave a question blank when you don't know the answer. Not on this test though.

When you take the SOL Chemistry test, you must fill in an answer to every single question, whether you know the answer or not.

Never Leave a Question Blank!

Just to prove the point, look at a case history. A pair of twins, Joe and Maria Bloggs, took the SOL Chemistry test last year. They had exactly the same knowledge of chemistry, and each of them knew the answer to exactly 25 out of the 60 questions. Joe filled in answers to the 25 questions he knew and left the other 25 blank. Maria, on the other hand knew that there was no guessing penalty, so she filled in the bubbles to the other 25 questions completely randomly. Out of the 25 guesses she made, she got 6 right. (If you know about probability, you know that you can expect to get about one out of every four random guesses right on this test).

So what happened? Well, the number of correct answers required to pass was 27, so Maria passed with a 31 (the 25 she knew plus the 6 she guessed right) and Joe failed with a 25. Maria passed and Joe failed even though they knew the exact same amount of chemistry! Once again, if you're going to play the game, you've got to know the rules.

THE TEST IS UNTIMED
There's absolutely no reason why you can't finish this test thoroughly, because you can't run out of time. You can go as slowly as you want, read as carefully as you want, and even close your eyes for a few minutes and take a little rest in the middle of the test.

WHAT YOU CAN USE DURING THE TEST
While you take the test, you will be able to use scratch paper, a four-function calculator, a ruler, and the periodic table of elements. Before you take the test, become familiar with the calculator you are going to use. You will be provided with a copy of the periodic table of elements when you take the test. Use it as much as you need to when answering test questions.

THE STATE OF VIRGINIA WANTS YOU TO DO WELL!
Think about taking two different tests, one given by someone who wants you do badly and one given by someone who wants you to do well. The person who wants you to badly will make the questions much more difficult, adding tricks and traps to try to make you choose wrong answers even if you know the material being tested. The person who wants you to do well will do everything possible to help you find the right answer. The people who write this test *don't* want you to fail. If you fail this test, it makes your school look bad and it makes Virginia look bad. That might seem like a lot of responsibility to heap on your shoulders when you're just trying to study chemistry, but it will work in your favor.

Some standardized tests, like the SAT, contain questions designed to trick the average test-taker into choosing wrong answers. This test is not like that. The people who write this test just want to see if you know the basics of chemistry, so they will make the questions as simple and straightforward as possible. There will be no tricks or traps, just ordinary multiple-choice questions about important chemistry topics.

WHERE TO START

There are several terrific ways that will help you prepare for this exam. At the top of the list has to be—bear with us for a moment here—studying. Not the "all nighter" type studying but good old fashioned "let's do a little review every day" kind of studying. While you're studying, don't forget to review your laboratory investigations as well. You will be specifically tested on your laboratory skills. With that said, assume that you've done all that and now the moment of truth has finally arrived and you're sitting in front of the SOL in Chemistry.

The proctor tells you to fill out all of the necessary information, passes out the rulers and calculators, reminds you to answer all 60 questions, and reminds you that the test is untimed. Now what? The proctor tells you to open the test booklet and begin.

TAKING THE TEST

You can start racking up points right away by reading through the questions first. Read *each word* of each question carefully. Missing one word in a question can change the meaning of the question or the "best answer" choice dramatically. Again, make sure that you bubble in your answer choice in the correct spot. And while we're on the topic of bubbling answers on the answer sheet, bubble *lightly* at first. If you press very hard when you bubble in and you need to change the answer sometime before you hand in the answer sheet, you run the risk of having the machine that grades the answer sheet pick up the answer that you did not intend.

All right, you've carefully read through each question on the exam and picked out and answered the easy questions. Now it's time to do a second pass through the exam and start spending time on those questions that you weren't sure of right off the bat. Here's where test-taking strategies come in handy.

STRATEGIES

USE PROCESS OF ELIMINATION (POE)

One of the answers for a multiple-choice question must be correct. If you don't know the answer, use Process of Elimination.

Say you read the following question:

1 What is the formula of a compound?
   A HCl
   B He
   C Cu
   D O_2

Now this might seem tricky but take a closer look. The question asks for a "compound," which means a combination of two or more elements. You can see that **B** and **C** can be ruled out because they are individual elements. By eliminating those choices, your chances are now 1 in 2. Choice **D** shows only one type of element as well but with a subscript of 2. Remember, you are looking for a combination of two or more **different** elements, so **D** can be ruled out. The correct answer is **A**.

YOU ARE LOOKING AT THE RIGHT ANSWER

The nicest thing about a multiple-choice test is that the correct answer is always right in front of you. That makes life much easier. It means that you don't have to come up with the correct answer from scratch. All you have to do is recognize it. Not only that, but sometimes you can also recognize wrong answers before you recognize the right one.

THE NAME GAME

Sometimes the choices themselves can help you eliminate them. See how this works.

What if you saw this question:

2 In which of the following processes is heat transferred in waves?
   F reduction
   G radiation
   H concentration
   J vaporization

Lesson 1: Test Structure and Strategies

Now this could really be a tough one to answer. All of answers end in "-ion." No help there. Check if the question can give you a better clue about how to look at the answers. If you go back to the question and put it into your own words, you see that they want you to find a process that moves heat from one place to another. So you can easily rule out **H**, which has to do with amounts not heat. Choice **F** is "reduction," which deals with the movement, or to be more specific, the gaining of electrons. Even if you didn't know the definition of reduction, you could probably deduce that the word "reduction" is not really associated with the transfer of heat. Now you are left with **G** and **J**. At this point, you could punt and choose one of the answers based on your gut reaction or you could try a little word association exercise.

Word association helps you figure out a word meaning through association with other words that sound similar. For instance, the definition of vaporization may not exactly leap out at you. Start to put together a list of words or phrases that sound similar, such as *vaporize,* or *vapor.* After a while, it may dawn on you that all of these words, including vaporization, have something to do with water, not heat. Do the same exercise for choice **G**, "radiation." There's *radiate*, which means to emit or spread out. There's *radiator*, which is a device that gives off heat. Hmmm, it looks like you've built a pretty strong case for choice **G**. That's your final answer, **G**.

Now you might be thinking, "I already know about word association." Well, that's great, you're already ahead of the game. But the thing to keep in mind is that word association is a tried and proven technique to help you score better. You shouldn't underestimate the value of this method. If you face a problem with similar sounding answers, take some time and try to associate words or phrases for each choice. You will have all the time you need for the test, so take advantage of it.

THE *YOU'VE GOT TO BE KIDDING* QUESTION
Try answering the following question:

3 The geometry of a molecule of SO_2 is—
   A linear
   B bent
   C triangular planar
   D triangular pyramidal

What if you have absolutely no clue? You should still guess. **Remember**, you should never leave any blanks on your answer sheet. You've got a one-out-of-four chance at getting any question right, even if you are completely clueless. The laws of probability say that some of your guesses will be right. The test makers will give you free points for your guesses, and you should take them up on their offer.

THAT'S NOT THE WAY IT'S DONE IN SCHOOL

When you do your homework or take a test in class, you usually show all your work. Your teacher looks at what you've done to make sure you understand the material. Sometimes, you'll get partial credit even though you didn't get the right answer. Other times, you may have the correct answer and still lose points because you didn't show how you arrived at it.

This test has different rules. You won't get any partial credit for almost knowing the answer, but you won't be penalized for taking a guess and getting the right answer. Remember, this test is like a game, and in order to win a game you need to play by the rules.

WHAT DOES THE TEST LOOK LIKE?

The Virginia Department of Education is very specific about the topics it expects you to know for the SOL chemistry exam. The exam has been broken down into five major reporting categories. The material covered in those categories are reviewed in the following lessons.

The test itself is 60 questions long. Of the 60 questions, only 50 will count toward your final score. The other 10 questions are field-test items. Because there is no way to know which 10 questions are field-test items, you need to treat each question on the exam as the real thing because for all intents and purposes, it is.

It's important to note that unlike standardized tests such as the PSAT and SAT, the EOC in Chemistry *does not* proceed along a significant order of increasing difficulty. In other words, the EOC in chemistry doesn't get harder as you move through the exam from beginning to end.

HOW IS THE CHEMISTRY EXAM SCORED?

As indicated earlier, a machine will score your answer sheet along with those of every chemistry student in Virginia. The machine automatically records your score on your answer sheet. Because the machine does the scoring, it's important that you do everything in your power to help the machine pick up the answer that you had intended. Remember to bubble in light at first in case you want to go back and change the answers to some of your questions. If you bubble in dark to start and have to erase an answer, the machine may pick up the answer that was not your final answer.

WRAPPING IT UP

These final tips may sound simple, but they are important. Prepare for the exam well in advance of test day. Eat well, sleep well, and relax.

GOOD LUCK!

LESSON 2
ATOMIC STRUCTURE

IT'S THE LITTLE THINGS—ATOMS

What is the smallest piece of a substance possible? If you take a piece of iron, for instance, and break it into smaller and smaller pieces, you will eventually have iron dust. These fine particles all have the properties of iron. You can, using special equipment, continue to break the iron into smaller and smaller pieces until you get iron atoms. The **atom** is the smallest particle of a substance that still retains the properties of that substance.

Atoms are made of three smaller particles: protons, neutrons, and electrons. The protons and neutrons are packed tightly together in the center of an atom creating the **nucleus**. The electrons, which are much smaller than the protons and neutrons, travel around the nucleus, kind of like planets orbiting the sun.

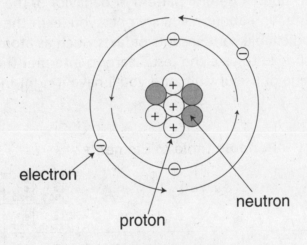

Atom Model

Protons: A proton is a positively charged particle in the nucleus of an atom. An atom gets its identity, or **atomic number**, from the number of protons in its nucleus. For instance, a helium atom *always* has 2 protons and its atomic number is 2. Likewise, a lithium atom *always* has 3 protons, and its atomic number is 3. If you change the number of protons in an element, you change the identity of the element too.

Neutrons: A neutron is an uncharged particle in the nucleus of an atom. A neutron is about the same size as a proton. While a particular element must always have the *same* number of protons, it can have *different* numbers of neutrons. For example, a boron atom must have 5 protons, but it can have 4, 5, 6, 7, or even 8 neutrons. Two boron atoms with differing numbers of neutrons are called **isotopes**. Isotopes are any atoms of the same element that have different numbers of neutrons.

Electrons: An electron is a negatively charged particle that revolves around the nucleus. An electron is much smaller than a proton or a neutron. Some electrons are very close to the nucleus, while others are much farther away. Atoms can give away electrons, gain them, or share them with other atoms. All chemical reactions are based upon the gain, loss, or sharing of electrons between atoms.

ELEMENTS AND THE PERIODIC TABLE

Elements are pure substances that cannot be broken down into simpler substances by physical or chemical means. Elements are made up of the same type of atoms, so the element iron is only made up of iron atoms. On Earth, there are about 100 naturally occurring elements There are also a number of man-made elements.

All of the elements are collected and organized on the **Periodic Table of Elements**. The periodic table is the most useful tool you have when you take a chemistry test, and you'll be given one to use when you take the Virginia SOL. The beauty of the periodic table is that you don't have to learn about each element individually. The periodic table is based upon a natural pattern of behavior of the elements, and so you can develop an intuition about reactivity once you learn the basic organization. In addition, the periodic table contains lots of facts, such as atomic mass and mass number, that you will need to take the test. Learn to interpret the periodic table because it's a gold mine of useful stuff, and you'll have it right in front of you when you take the test!

Periodic Table of Elements

1 H 1.0																	2 He 4.0
3 Li 6.9	4 Be 9.0											5 B 10.8	6 C 12.0	7 N 14.0	8 O 16.0	9 F 19.0	10 Ne 20.2
11 Na 23.0	12 Mg 24.3											13 Al 27.0	14 Si 28.1	15 P 31.0	16 S 32.1	17 Cl 35.5	18 Ar 39.9
19 K 39.1	20 Ca 40.1	21 Sc 45.0	22 Ti 47.9	23 V 50.9	24 Cr 52.0	25 Mn 54.9	26 Fe 55.8	27 Co 58.9	28 Ni 58.7	29 Cu 63.5	30 Zn 65.4	31 Ga 69.7	32 Ge 72.6	33 As 74.9	34 Se 79.0	35 Br 79.9	36 Kr 83.8
37 Rb 85.5	38 Sr 87.6	39 Y 88.9	40 Zr 91.2	41 Nb 92.9	42 Mo 95.9	43 Te (98)	44 Ru 101.	45 Rh 102.	46 Pd 106.4	47 Ag 107.	48 Cd 112.	49 In 114.8	50 Sn 118.	51 Sb 121.	52 Te 127.6	53 I 126.	54 Xe 131.
55 Cs 132.	56 Ba 137.3	57 *La 138.	72 Hf 178.5	73 Ta 180.	74 W 183.9	75 Re 186.	76 Os 190.2	77 Ir 192.	78 Pt 195.1	79 Au 197.	80 Hg 200.6	81 Tl 204.4	82 Pb 207.	83 Bi 209.0	84 Po (209)	85 At (210)	86 Rn (222)
87 Fr (223)	88 Ra 226.0	89 †Ac 227.	104 Unq (261)	105 Unp (262)	106 Unh (263)	107 Uns (262)	108 Uno (265)	109 Une (267)									

*Lanthanide

58 Ce 140.	59 Pr 140.9	60 Nd 144.	61 Pm (145	62 Sm 150.	63 Eu 152.	64 Gd 157.3	65 Tb 158.	66 Dy 162.	67 Ho 164.9	68 Er 167.	69 Tm 168.	70 Yb 173.0	71 Lu 175.

†Actinide

90 Th 232.0	91 Pa (231)	92 U 238.0	93 Np (237)	94 Pu (244)	95 Am (243)	96 Cm (247)	97 Bk (247)	98 Cr (251)	99 Es (252)	100 Fm (257)	101 Md (258)	102 No (259)	103 Lr (260)

On the periodic table, elements are arranged in horizontal rows, called **periods,** and vertical columns called **groups**. Starting with the lowest atomic number element, hydrogen, in the upper-left corner, the atomic number increases going across the row to helium. The atomic number increases in the next row, starting with lithium, and increases to the right. Each element is represented on the table with its symbol, atomic number, and atomic weight.

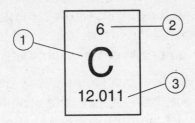

① This is the **symbol** for the element carbon. Every element has a one-, two-, or three-letter symbol that appears on the periodic table. Most symbols come from the first letter of the element's name. But if two elements start with the same letter, say carbon and calcium, the element with the lowest atomic number usually uses only the first letter. In this case, C represents carbon (atomic number 6), and not calcium (atomic number 20), which uses its first two letters, Ca, for the symbol. Sometimes, the symbol comes from the Latin name. For example, sodium's symbol is Na (Latin name: natrium). The good news is that you don't have too many of these to worry about, and the Virginia SOL isn't going to try to trick you.

② This is the **atomic number** of the element. The atomic number represents the number of protons in the nucleus of an element; it is also equal to the number of electrons surrounding the nucleus of an element in its neutral state. The periodic table shows that the atomic number of carbon is 6. A carbon atom must *always* have 6 protons in its nucleus.

③ This is the **atomic weight** of the element. The atomic weight given on the periodic table is the *average of the mass numbers of a large sample of isotopes of an element.*

Lesson 2: Atomic Structure

You know that the nucleus of the atom is made up of protons and neutrons. The **mass number** of an atom is the total number of its neutrons and protons. (Electrons do not add to the mass, because they are so small.) So if a carbon atom has 6 protons and 6 neutrons, then its mass number is 12. However, this is *not* its atomic weight. Unlike the number of protons, the number of neutrons in an element *can* vary. In fact, most elements are known to have some atoms with differing numbers of neutrons. That is, some of the carbon atoms will have 6 protons and 6 neutrons (atomic mass = 12), but others might have 6 protons and 8 neutrons (atomic mass = 14). Remember, atoms of an element with differing numbers of neutrons are called **isotopes**. Isotopes of the same element have the same number of protons, but different numbers of neutrons.

The atomic weight of an element will give you a pretty good idea of the most common isotope of that element. For instance, the atomic weight of carbon is 12.011. That's awfully close to 12, and it turns out that about 99% of the carbon in existence is the isotope carbon-12.

Sometimes, an atom is represented by a symbol with superscripts and subscripts on its left, showing the atom's mass number and atomic number, respectively. The symbol for boron with 5 protons and 6 neutrons is shown below.

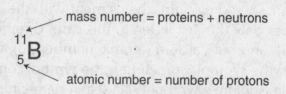

Remember, the **atomic mass** is given for a single isotope of an element. The **atomic weight** given on the periodic table is the average of the mass numbers of a large sample of isotopes of an element.

ELECTRONS

In an element, electrons are where the action is. All chemical reactions arise from the gain, loss, or sharing of electrons. Although an atom can have anywhere from one to more than a hundred electrons, the electrons are always arranged around the nucleus in an orderly and predictable way. Because the periodic table is based upon the number and arrangement of electrons in each element, you can tell a lot about an element's reactivity from its place in the periodic table.

Even though an element can have dozens of electrons, thankfully, you don't have to keep up with all of them because only the outermost electrons are involved in reactions. These reactive electrons are called **valence electrons**. *The valence electrons are the most important electrons in an atom because they form bonds with other atoms.*

SHELLS

In an atom, the electrons are grouped into **shells**. Each shell has a certain distance from the nucleus and energy level. Electrons in the first shell are closest to the nucleus and electrons in higher numbered shells are farther away. Also, electrons in higher numbered shells have more energy than electrons in lower numbered shells.

The larger the shell number, the more electrons it can hold. The first shell can hold only 2 electrons, the second shell can hold 8, the third shell can hold 18, and so on. A really important point is that, even though the higher numbered shells can hold more than 8 electrons, they will usually manage to arrange them into subshells where *there are 8 valence electrons*. So no matter how big the shell gets, you can think of an atom below the first row as having a valence shell that can hold 8 electrons.

This brings us to one of the most fundamental ideas of chemistry: the **octet rule**. Specifically, every atom is driven to have a full shell of valence electrons, or a **full octet**. For example, if an element has 6 valence electrons, it will form bonds and share electrons so that it has 8 valence electrons. No more, no less. Eight.

There are exceptions to this rule, the most notable being the first row elements, hydrogen and helium. These elements have a shell that can hold 2 electrons, so they will have a maximum of 2 valence electrons. There are also elements below the second row that can hold more than 8 electrons, but you will not be expected to know them. Do expect most elements to follow the octet rule.

The horizontal rows of the periodic table are called **periods**, and as we move down the table, each period corresponds to the addition of a new shell. All the elements in the same period will have their valence electrons in the same shell. For instance, the elements in the third period (Na, Mg, Al, Si, P, S, Cl, and Ar) all have valence electrons in the third shell. Notice that each element has a different number of valence electrons, increasing as you go across the period. *This is important because these differences in valence electrons are the basis of the different chemical personalities of each element.* Why is that? Because elements with different numbers of valence electrons will need to bond differently in order to gain a full octet.

The vertical columns of the periodic table are called **groups**, and the elements in a group all contain the same number of valence electrons. For instance, the elements in the second group (Be, Mg, Ca, Sr, Ba, and Ra) all have 2 valence electrons in their outer shells. *Because they have the same number of valence electrons, the elements in a group will have similar chemical personalities.* Why? Because elements with the same number of valence electrons will bond similarly in order to fill their octets. Hopefully, this gives you a glimpse of how the periodic table can help you.

SUBSHELLS

Each shell can be further divided into subshells. An electron's subshell describes the shape of the electron cloud.

The first shell has one subshell, called the *s* subshell.

The second shell has two subshells, an *s* subshell and a *p* subshell.

The third shell has three subshells, *s*, *p*, and *d*.

Higher numbered shells all have four subshells, *s*, *p*, *d*, and *f*.

The *s* subshell can hold only 2 electrons. Electrons in the *s* subshells form spherical clouds.

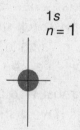

The *p* subshell can hold 6 electrons. Electrons in the *p* subshells form dumbbell-shaped clouds.

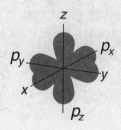

The *d* subshell can hold 10 electrons and the *f* subshell can hold 14 electrons. Their shapes are too complicated to describe here, but you don't need to worry about them.

Within each subshell, electrons can be grouped in **orbitals**. No matter what subshell, there can only be 2 electrons in each orbital. So the 2 electrons in the *s* subshell, are placed in a single orbital, while the 6 electrons in the *p* subshell are divided among 3 orbitals. The two electrons in an orbital are distinguished by their **spin**. When there are two electrons in an orbital, one is said to have positive spin and the other has negative spin.

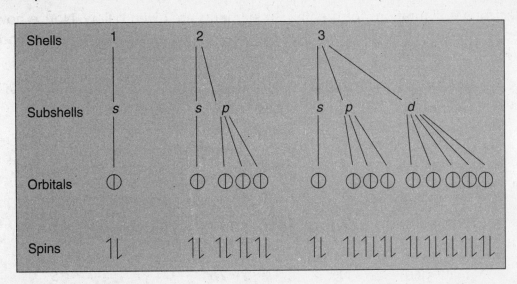

Electron Configurations

Every atom has its electrons arranged into orbitals within shells and subshells. This is an atom's **electron configuration**. An atom's electron configuration tells us where all the electrons are within the shells and subshells. To start, identify the number of electrons in an atom. A neutral atom will have the same number of electrons as protons, so you can tell the number of electrons in an atom by looking at the atomic number. For example, a neutral nitrogen atom has the atomic number 7, so it has 7 electrons. Electrons are always placed in the shells and subshells with the lowest possible energy. This is like filling a glass in which water fills from the bottom to the top: low-energy shells are filled with electrons first, then electrons gradually fill shells with higher energy.

The electron configuration of a nitrogen atom is shown below.

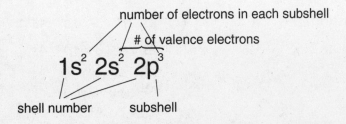

Lesson 2: Atomic Structure

In the electron configuration for the nitrogen atom above, you can see that there are 2 electrons in the 1s subshell, 2 electrons in the 2s subshell, and 3 electrons in the 2p subshell. Remember that the number of electrons in the subshells (2 + 2 + 3) must always add up to the total number of electrons in the atom (7). The outermost shell is the 2 shell, so the valence electrons are all the electrons in the 2 subshells (2 + 3) or 5.

The electron configurations for the first 20 elements are given below. Notice how each added electron goes into the lowest-energy shell and subshell available and how new shells and subshells are started when lower-energy ones are filled.

Atomic #	Element	Symbol	Electron Config.	# of Valence Electrons
1	Hydrogen	H	$1s^1$	1
2	Helium	He	$1s^2$	2 or zero
3	Lithium	Li	$1s^2\ 2s^1$	1
4	Beryllium	Be	$1s^2\ 2s^2$	2
5	Boron	B	$1s^2\ 2s^2 2p^1$	3
6	Carbon	C	$1s^2\ 2s^2 2p^2$	4
7	Nitrogen	N	$1s^2\ 2s^2 2p^3$	5
8	Oxygen	O	$1s^2\ 2s^2 2p^4$	6
9	Fluorine	F	$1s^2\ 2s^2 2p^5$	7
10	Neon	Ne	$1s^2\ 2s^2 2p^6$	8 or zero
11	Sodium	Na	$1s^2\ 2s^2 2p^6\ 3s^1$	1
12	Magnesium	Mg	$1s^2\ 2s^2 2p^6\ 3s^2$	2
13	Aluminum	Al	$1s^2\ 2s^2 2p^6\ 3s^2 3p^1$	13
14	Silicon	Si	$1s^2\ 2s^2 2p^6\ 3s^2 3p^2$	4
15	Phosphorous	P	$1s^2\ 2s^2 2p^6\ 3s^2 3p^3$	5
16	Sulfur	S	$1s^2\ 2s^2 2p^6\ 3s^2 3p^4$	6
17	Chlorine	Cl	$1s^2\ 2s^2 2p^6\ 3s^2 3p^5$	7
18	Argon	Ar	$1s^2\ 2s^2 2p^6\ 3s^2 3p^6$	8 or zero
19	Potassium	K	$1s^2\ 2s^2 2p^6\ 3s^2 3p^6\ 4s^1$	
20	Calcium	Ca	$1s^2\ 2s^2 2p^6\ 3s^2 3p^6\ 4s^2$	

Of course, you can also use the periodic table to tell the electron configuration of any element. The order in which shells and subshells are filled can be seen by following the table from left to right across each period. The elements are arranged in the same way orbitals are filled: First the 1s shell gets 2 electrons, then the 2s subshell holds two more, then the 2p subshell has 6 for a total of 10 electrons.

1s					1s
2s				2p	
3s				3p	
4s		3d		4p	
5s		4d		5p	
6s	4f	5d		6p	
7s	5f	6d			

4f
5f

You should note that after the third period, the filling of subshells becomes more complicated. Notice, for instance, that the 4s subshell fills before the 3d subshell, then finally the 4p subshell. Because of the different shapes of the orbitals, the 4s subshell has a lower energy level than the 3d subshell. You don't need to worry about the actual shapes, but should know that as we fill orbitals, the 4s comes before 3d and then the 4p last. This is easy to remember because the atomic numbers increase in the same order as the subshells are filled: 4s, 3d, 4p.

Just a Little History

Our current understanding of the structure of the atom has been developing for thousands of years, with most of the development occurring in the last two hundred. It's still developing, by the way, and there are still many things we don't understand about atoms.

About 2,500 years ago, the Greek philosopher **Democritus** came up with the idea that all matter is composed of tiny, indivisible particles that he called atoms. After that, not much happened in the field of atomic theory for more than 2,000 years.

In the early 1800s, **John Dalton** was the first to put forth that there are many different kinds of atoms, which he called elements. He said that elements combine to form compounds in precise ratios. Water (H_2O), for instance, always has two hydrogen atoms for every oxygen atom. He also said that atoms are never created or destroyed in chemical reactions.

Lesson 2: Atomic Structure

By the end of the 1800s, scientists had identified many of the most common elements. They had even started arranging them in an early version of the periodic table, but they still didn't know what atoms were made of. In the late 1800s, **J.J. Thomson** put forth the idea that atoms were composed of positive and negative charges. The negative charges were called electrons, and Thomson guessed that they were sprinkled throughout the positively charged atom like chocolate chips sprinkled throughout a blob of cookie dough.

In the early 1900s, **Ernest Rutherford** discovered that all of the positive charge in an atom was concentrated in the center, and that an atom is mostly empty space. This led to the idea that an atom has a positively charged nucleus, which contains most of an atom's mass, and that the tiny, negatively charged electrons travel around this nucleus.

Around this time, **Max Planck** figured out that energy is quantized. That means that energy increases occur in steps, not on a smooth curve. That means that an atom's electrons can't travel all over the place. Instead, each electron must stay in a certain region around the nucleus, depending on the electron's energy.

Neils Bohr took the quantum theory and used it to predict that electrons orbit the nucleus at specific, fixed distances like planets orbiting the sun. We still use this model today, even though actual atoms are more complicated than that.

Atoms turn out to be more complicated than planets. **Werner Heisenberg** said that it is impossible to know both the position and momentum of an electron at a particular instant. In terms of atomic structure, this means that electrons don't travel in specific paths, like those of planets. Instead, an electron's shell, subshell, and orbital describe the probability that an electron will be found in some general region.

Now that you've learned some basic features of the periodic table and atoms, try using what you know to answer some questions. Remember to use POE to help you.

REVIEW FOR LESSON 2

1. What is the chemical symbol for neon?

   A N

   B Na

   C Ne

   D Ni

2. How many protons are in a fluorine atom?

   F 9

   G 10

   H 19

   J 28

3. How many protons are in a beryllium atom?

   A 4

   B 5

   C 9

   D 13

4. The number 32, in phosphorous-32 refers to the—

   F number of protons in the atom

   G number of neutrons in the atom

   H total number of protons and neutrons in the atom

   J total number of electrons and protons in the atom

5. An atom of aluminum-27 contains—

   A 27 protons and 27 neutrons

   B 13 protons and 27 neutrons

   C 13 protons and 14 neutrons

   D 13 protons and 13 neutrons

6 An atom of argon-40 contains—
   F 18 protons and 18 neutrons
   G 18 protons and 22 neutrons
   H 18 protons and 40 neutrons
   J 20 protons and 20 neutrons

7 Which of the following atoms contains 11 protons and 12 neutrons?
   A $^{12}_{11}Na$
   B $^{23}_{11}Na$
   C $^{23}_{12}Mg$
   D $^{24}_{12}Mg$

8 Which of the following atoms contains 33 protons and 42 neutrons?
   F $^{75}_{33}As$
   G $^{75}_{42}As$
   H $^{75}_{42}Mo$
   J $^{96}_{42}Mo$

9 The atomic weight of an element is—
   A the number of protons in the element
   B the number of neutrons in the element
   C the weight of the heaviest isotope of the element
   D the average weight of all the isotopes of the element

10 The element boron has only two stable isotopes. One stable isotope has a mass number of 10 and the other has a mass number of 11. Which of the following could be the atomic weight of the element?
   F 9.5
   G 10.8
   H 11.7
   J 12.4

11 An atom's valence electrons are the electrons in its—
   A innermost shell
   B outermost shell
   C nucleus
   D spin state

12 An atom's behavior in making bonds is determined mainly by its—
   F total number of neutrons
   G total number of electrons
   H total number of protons
   J number of valence electrons

13 Which of the following atoms has 7 valence electrons?
   A Fluorine
   B Oxygen
   C Nitrogen
   D Carbon

14 What is the electron configuration for a sodium atom?
   F $1s^2\ 2s^22p^1$
   G $1s^2\ 2s^22p^4$
   H $1s^2\ 2s^22p^6\ 3s^1$
   J $1s^2\ 2s^22p^6\ 3s^23p^5$

15 Which of the following atoms has valence electrons in the $2p$ subshell?
   A Be
   B N
   C Mg
   D Cl

Review for Lesson 2

16. Which neutral atom has the following electron configuration?

$$1s^2 \quad 2s^2 2p^6 \quad 3s^2 3p^6$$

F Neon
G Magnesium
H Sulfur
J Argon

17. How many valence electrons are there in a magnesium atom?
A 1
B 2
C 3
D 4

18. How many valence electrons are there in an aluminum atom?
F 1
G 2
H 3
J 4

19. The mass of an electron is—
A greater than the mass of a proton
B the same as the mass of a proton
C the same as the mass of a neutron
D less than the mass of a neutron

20. How many valence electrons are there in a carbon atom?
F 1
G 2
H 4
J 6

ANSWERS AND EXPLANATIONS

1. **C is correct.** Ne is the symbol for neon. N = nitrogen, Na = sodium, Ni = nickel. If you weren't sure of the right answer, Ni and Na could be ruled out as unlikely simply because of the spelling. This would leave only two choices, giving you a 50 percent chance of guessing the correct answer.

2. **F is correct.** Fluorine's atomic number, 9, tells the number of protons in its nucleus. If you don't know from the periodic table how to find the atomic number, then just look at the symbol for fluorine. Only two numbers are given, 9 and 19. Most likely one of these will be the right answer, so POE once again will increase your chances when guessing.

3. **A is correct.** Beryllium's atomic number, 4, tells the number of protons in its nucleus.

4. **H is correct.** The number at the end of an isotope tells you the total number of protons and neutrons in the atom. By looking at the symbol for phosphorous, P, in the periodic table we know the atomic number is 15 and the atomic weight 31. So choices **F** and **J** can be ruled out.

5. **C is correct.** The number of protons in an atom is given by the atomic number (13 in this case). **You can get the number of neutrons by subtracting the atomic number from the mass number** ($27 - 13 = 14$). If all you know is that there are 13 protons in aluminum, you can still use POE here to eliminate choice **A**.

6. **G is correct.** The number of protons in an atom is given by the atomic number (18 in this case). You can get the number of neutrons by subtracting the atomic number from the mass number ($40 - 18 = 22$).

7. **B is correct.** $^{23}_{11}$Na has 11 protons (atomic number) and 12 neutrons (mass number minus atomic number). If you know that Na has 11 protons, you can use POE to eliminate choices **C** and **D**.

8. **F is correct.** $^{75}_{33}$As has 33 protons (atomic number) and 42 neutrons (mass number minus atomic number). Only answer choice **F** shows the correct atomic number.

9. **D is correct.** The atomic weight is the average of the mass numbers of a large sample of isotopes of an element.

10. **B is correct.** The atomic weight is the average of the mass numbers of all the isotopes. If the mass numbers of the isotopes are 10 and 11, then the atomic weight must be somewhere in between 10 and 11.

Review for Lesson 2

11 **B is correct.** Valence electrons are the high energy electrons in an atom's outermost shell. You may not know where the valence electrons are, but they are probably not in the nucleus, so you can use POE to eliminate choice **C**.

12 **J is correct.** An atom will form bonds in an effort to get a complete octet of electrons, so the number of valence electrons determines the kinds of bonds an atom must make in order to complete its outer shell. If you only know that bonding depends on electrons, you can still eliminate choices F and H.

13 **A is correct.** Fluorine has the following electron configuration: $1s^2\ 2s^22p^5$. It has 7 electrons in its outer shell. You can also tell by looking at the periodic table.

14 **H is correct.** A quick way to get this one is to remember that the atomic number gives the number of electrons (when it is not bonded to another atom), as well as the number of protons. Sodium has an atomic number of 11, and choice **H** is the only one whose electrons numbers add up to 11 (2 + 2 + 6 + 1 = 11).

15 **B is correct.** If you look at the periodic table, you can see that N is in the section whose valence electrons are in the $2p$ subshell. Chlorine has valence electrons in the $3p$ subshell, Beryllium has valence electrons in the $2s$ subshell, and Magnesium has valence electrons in the $3s$ subshell.

16 **J is correct.** Argon has the electron configuration $1s^2\ 2s^22p^6\ 3s^23p^6$. The electrons add up to 18, which is the atomic number of Ar. Notice that argon has all of its valence shells filled. Because it has a complete octet, it doesn't need to form bonds with other atoms.

17 **B is correct.** Magnesium has only 2 electrons in its third shell, both in the $3s$ subshell.

18 **H is correct.** Aluminum has 3 valence electrons in its third shell. It has 2 electrons in the $3s$ subshell and 1 in the $3p$ subshell.

19 **D is correct.** Protons and neutrons have the same mass, and an electron's mass is much smaller than either.

20 **H is correct.** Carbon has 4 valence electrons in its second shell, 2 in the $2s$ subshell and 2 in the $2p$ subshell.

Note: Most students don't memorize electron configuration. They learn how many electrons are in the outer shell of a group.

LESSON 3
THE PERIODIC TABLE

METALS AND NONMETALS

The elements on the periodic table can be divided neatly into groups: metals and nonmetals. A **metal** is an element that usually loses electrons in a bond. A **nonmetal** is an element that usually gains electrons in a bond. **Metalloids** are in between the metals and nonmetals and can behave a little bit like both. The metals, nonmetals, and metalloids are shown on the periodic table below.

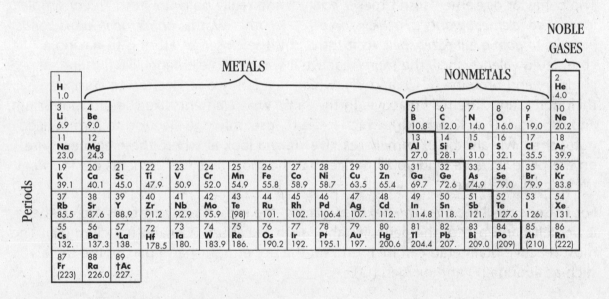

You can see that **nonmetals** are close to the right-hand side of the periodic table. Nonmetals are only a few electrons short of having a complete octet in their valence shells, so it's easier for them to complete their valence shells by gaining a few electrons rather than by losing several. In their elemental forms, nonmetals can be found in nature as gases, liquids, or crystalline solids. Think of some examples you know: oxygen (O_2) is a gas, carbon is found as diamond or graphite, and bromine is a liquid.

Looking at the rest of the periodic table, you can see that the vast majority of the elements are **metals**. Metals only have a few electrons in their valence shells, so it is easier for them to have filled outer shells by losing electrons. In their elemental forms, metals tend to be hard, shiny solids that are good conductors of heat and electricity. Gold, silver, and iron are metals.

The **metalloids,** or semimetals, are elements along the step-like line that separates metals and nonmetals. Metalloids have personality traits like both metals and nonmetals, and can both look and act like either. For example, metalloids can fill their octets by either giving up or gaining electrons in a bond and can be fairly good conductors of electricity (that's why they are sometimes called **semiconductors**). Also, they can be either amorphous or metal-like solids. Some common examples of metalloids are Si (silicon) or Sb (antimony).

In contrast to all the reactivity found in the rest of the table, in the far right column of the periodic table are the **noble gases**. These elements already *have* a full octet and are notable for their nearly complete lack of reactivity.

The behavior of elements isn't rocket science and really boils down to a very simple idea: *Every element wants to achieve an octet.* In other words, *an element forms bonds in order to gain a full outer shell.* As it turns out, the closer an atom is to having a completed valence shell, the more reactive it will be. So elements like fluorine, chlorine, and bromine (F, Cl, and Br), which need only one electron to complete their octet, are extremely reactive. In the same way, elements like sodium, potassium, and rubidium (Na, K, and Rb), which need to lose only one electron to form a filled outer shell, will also be extremely reactive. Take a look at where these elements are on the periodic table, and you'll notice that the most reactive elements are found at the sides of the table.

For reasons we'll see over the next few pages, the most reactive metals are on the lower left-hand part of the periodic table, such as cesium (Cs) and francium (Fr). The most reactive nonmetals can be found on the upper right-hand part of the table, such as fluorine (F) and oxygen (O).

PERIODIC TRENDS AND REACTIVITY

THE PERIODIC TABLE

You can make very good guesses about the reactivity of an element by just knowing where the element is found on the periodic table. Most of the work has been done for you because the periodic table organizes elements in a way that tells you about their chemical personalities. Some of the traits that affect an atom's bonding behavior are its size, its ability to hang onto its electrons, and its ability to take electrons away from other atoms. As you look across or up and down the periodic table, you will notice that these traits follow predictable **periodic trends**.

Look at the trends that will be most useful for you on the Virginia SOL:

ATOMIC RADIUS

The size of an atom relates to the electron cloud around the nucleus. Although electrons are small, they fill up a lot of space because they are always moving. The electrons farthest from the nucleus are the valence electrons. The **atomic radius** is the distance from the nucleus of an atom to its valence electrons. Basically, the atomic radius tells the size of the atom.

Periodic Table of Elements

1 H 1.0																	2 He 4.0
3 Li 6.9	4 Be 9.0											5 B 10.8	6 C 12.0	7 N 14.0	8 O 16.0	9 F 19.0	10 Ne 20.2
11 Na 23.0	12 Mg 24.3											13 Al 27.0	14 Si 28.1	15 P 31.0	16 S 32.1	17 Cl 35.5	18 Ar 39.9
19 K 39.1	20 Ca 40.1	21 Sc 45.0	22 Ti 47.9	23 V 50.9	24 Cr 52.0	25 Mn 54.9	26 Fe 55.8	27 Co 58.9	28 Ni 58.7	29 Cu 63.5	30 Zn 65.4	31 Ga 69.7	32 Ge 72.6	33 As 74.9	34 Se 79.0	35 Br 79.9	36 Kr 83.8
37 Rb 85.5	38 Sr 87.6	39 Y 88.9	40 Zr 91.2	41 Nb 92.9	42 Mo 95.9	43 Tc (98)	44 Ru 101.	45 Rh 102.	46 Pd 106.4	47 Ag 107.	48 Cd 112.	49 In 114.8	50 Sn 118.	51 Sb 121.	52 Te 127.6	53 I 126.	54 Xe 131.
55 Cs 132.	56 Ba 137.3	57 *La 138.	72 Hf 178.5	73 Ta 180.	74 W 183.9	75 Re 186.	76 Os 190.2	77 Ir 192.	78 Pt 195.1	79 Au 197.	80 Hg 200.6	81 Tl 204.4	82 Pb 207.	83 Bi 209.0	84 Po (209)	85 At (210)	86 Rn (222)
87 Fr (223)	88 Ra 226.0	89 †Ac 227.	104 Unq (261)	105 Unp (262)	106 Unh (263)	107 Uns (262)	108 Uno (265)	109 Une (267)									

*Lanthanide

58 Ce 140.	59 Pr 140.9	60 Nd 144.	61 Pm (145	62 Sm 150.	63 Eu 152.	64 Gd 157.3	65 Tb 158.	66 Dy 162.	67 Ho 164.9	68 Er 167.	69 Tm 168.	70 Yb 173.0	71 Lu 175.

†Actinide

90 Th 232.0	91 Pa (231)	92 U 238.0	93 Np (237)	94 Pu (244)	95 Am (243)	96 Cm (247)	97 Bk (247)	98 Cf (251)	99 Es (252)	100 Fm (257)	101 Md (258)	102 No (259)	103 Lr (260)

Moving from Left to Right Across a Period (Li to Ne, for Instance), Atomic Radius Decreases

This might seem to contradict logic because the number of electrons is *increasing* as you move across a period. But think of it this way: moving from left to right across a period, protons are added to the nucleus and, therefore, the valence electrons feel a greater positive charge. The increasing nuclear charge pulls the electrons closer to the protons, which *decreases* the atomic radius. You might be thinking, "But electrons are also being added, so don't they cancel out the effect of an increasing number of protons?" This is a great question, and the answer is no, the electrons don't cancel out the effect of the positive proton charges. Remember that the electrons are being added into the same shell and so are about an equal distance from the nucleus. Because of this, *electrons within a shell all feel the same pull*. The electrons are not shielded from the nuclear charge so atomic radius decreases as we move left to right. **Electron shielding** becomes important when we talk about effects of electrons from different shells.

Moving Down a Group (Li to Cs, for Instance), Atomic Radius Increases

Moving down a group, shells of electrons are added to the nucleus, and shielding is now important. Each shell shields the more distant shells from the nucleus and cancels some of the positive charge felt by the outer electrons. As you would expect, the valence electrons get farther and farther away from the nucleus, and so the atomic radius increases as you move down a group. Remember, electrons within a shell do not shield each other, but electrons in different shells do.

IONIZATION ENERGY

Electrons are attracted to the nucleus of an atom, so it takes energy to remove an electron. The energy required to remove an electron from an atom is called the **first ionization energy**. Removing an electron creates a charge imbalance and so a positive ion, or cation, is formed. The energy required to remove the next electron from the ion is called the **second ionization energy**, and so on.

Moving from Left to Right Across a Period, Ionization Energy Increases

You can use the same arguments for atomic radius to explain the trend in ionization energy. Moving from left to right across a period, protons are added to the nucleus, which increases its positive charge. Because electrons within a shell don't shield each other, the valence electrons experience an increasing positive nuclear charge, so it takes more energy to remove them. You can also look at this by using what you know about the atomic radius: as we move across the period, atomic radius decreases because the electron cloud contracts with increasing nuclear charge. Because the electrons are closer to the nucleus, they are more tightly bound and, therefore, require more energy to remove.

Moving Down a Group, Ionization Energy Decreases

This trend makes sense given all you know about shielding and atomic radius. Moving down a group, shells of electrons are added, and each inner shell shields outer shells from the nucleus. As a result, the outermost valence electrons feel less and less nuclear charge as you go down a group, and so it is easier to remove an electron. Think about it from the standpoint of atomic radius. Atomic radius increases going down a group. Because these electrons are further from the protons, it's easier to remove them. That explains why ionization energy decreases down a group.

The Second Ionization Energy Is Greater Than the First Ionization Energy, and So On

When an electron has been removed from an atom, the number of protons becomes greater than the number of electrons and, as you would think, the remaining valence electrons move closer to the nucleus. This increases the attractive force between the electrons and the nucleus, increasing the second ionization energy. In other words, it's harder to remove the second electron than the first.

As Electrons Are Removed, Ionization Energy Increases Gradually until a Shell Is Empty, Then It Makes a Big Jump

For each element, when the valence shell is empty, the next electron must come from a shell that is much closer to the nucleus, making the ionization energy for that electron much larger than for the previous ones. Look at some examples:

For Na (in Group 1), the second ionization energy is much larger than the first.

For Mg (in Group 2), the first and second ionization energies are comparable, but the third is much larger than the second.

For Al (in Group 3), the first three ionization energies are comparable, but the fourth is much larger than the third.

Electron Affinity

Moving from Left to Right Across a Period, Electron Affinity Increases

To have an affinity means to like something. Electron affinity tells us how much an element likes to take on an extra electron by measuring the energy change when an electron is added. When the addition of an electron makes the atom more stable, energy is given off. This is true for most of the elements. When the addition of an electron makes the atom less stable, energy must be put in. This happens when the added electron must be placed in a higher energy level, which decreases stability. This is the case for elements with full subshells, such as the alkaline earths and the noble gases.

Moving from left to right across a period, the energy given off when an electron is added increases. Electron affinities don't change very much moving down a group.

Electronegativity

Electronegativity is thought of in terms of two atoms in a bond and refers to how strongly the nucleus of an atom attracts the electrons of other atoms in a bond. For example, in the molecule HF, fluorine and hydrogen share an electron pair in a bond. But the fluorine atom attracts more than its fair share of the electron pair, and we say that fluorine is more electronegative than hydrogen. Electronegativities of elements are estimated based on ionization energies and electron affinities and follow basically the same trends. In general, the nonmetals, which gain electrons in bonds, have higher electronegativities than metals, which lose electrons in bonds.

Moving from Left to Right Across a Period, Electronegativity Increases

Moving across a period, atomic radius decreases and the electrons move closer to the nucleus. Another way to look at this is that *the energy of the valence electron shell decreases*, or becomes more stable, and, therefore, is better able to take on an extra electron. If you were an electron, wouldn't you want to be in a more stable shell? Electronegative elements attract electrons because they can best stabilize them. The shell stability increases from left to right within a period, and, therefore, the electronegativity increases as well.

Moving Down a Group, Electronegativity Decreases

As electron shells are added, shielding reduces the attraction of an atom's nucleus to the valence shell. As outer shells are added, the shell stability decreases and is less able to stabilize added electrons. Therefore, electronegativity decreases as we move down a group.

Think about all you know from shielding, full octets, and the trends for atomic radius and ionization, and electronegativity makes sense.

Review

Before moving on, look at all the things that happen as you move across a period. Protons are added to the nucleus and electrons are added to the valence shell. Because the electrons in the valence shell do not shield one another, they feel an increasing positive nuclear charge.

- Atomic radius decreases.

- Ionization energy increases.

- Electron affinity increases.

- Electronegativity increases.

The various periodic trends are summarized in the diagram below.

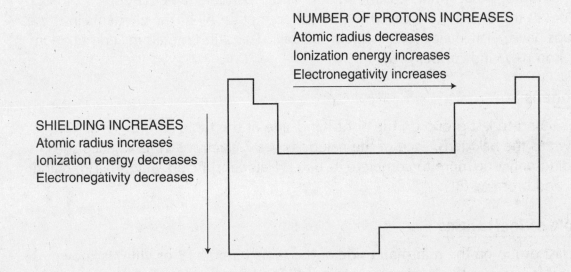

Lesson 3: The Periodic Table

GROUP NAMES

Certain vertical groups on the periodic table are known by specific names that you should know.

Alkali Metals

The first group on the left-hand side of the table (Group 1 or IA) is called the alkali metals. Each of the alkali metals has exactly one valence electron. Alkali metals include lithium (Li), sodium (Na), and potassium (K).

Alkaline Earth Metals

The second group on the left-hand side of the table (Group 2 or IIA) is called the alkaline earth metals. Each of the alkaline earth metals has exactly two valence electrons. Alkaline earth metals include beryllium (Be), magnesium (Mg), and calcium (Ca).

Transition Metals

The ten groups that run across the middle of the periodic table (Groups 3 through 12, or IIIB through IIB) are called the transition metals. All of the elements in these groups have valence electrons in the d subshell. Transition metals include titanium (Ti), iron (Fe), and copper (Cu).

Halogens

The second-to-last group on the right-hand side of the table (Group 17 or VIIA) is known as the halogens. Each of the halogens has 7 electrons in its valence shell, needing only one more to complete its octet. Halogens include fluorine (F), chlorine (Cl), and bromine (Br).

Noble or Inert Gases

The last group on the right-hand side of the table (Group 18 or VIIIA) is known as the noble gases or the inert gases. Each of the noble gases has 8 electrons in its valence shell, so noble gases are extremely unreactive because they already have complete octets. Noble gases include helium (He, whose valence shell is complete with only 2 electrons), neon (Ne), and argon (Ar).

REVIEW FOR LESSON 3

1. Which of the following elements is a nonmetal?

   A Magnesium
   B Iron
   C Sulfur
   D Lithium

2. Which of the following elements is a metal?

   F Phosphorous
   G Sodium
   H Neon
   J Oxygen

3. Which of the following elements has properties of both metals and nonmetals?

   A As
   B F
   C Ni
   D Ba

4. Which of the following elements is the most reactive?

   F Rb
   G Ca
   H Mg
   J K

5. Which of the following elements is the most reactive?

   A S
   B Se
   C P
   D Cl

6. Which of the following describes how difficult it is to remove a valence electron from an atom of a particular element?

   F Electronegativity
   G Ionization energy
   H Electron affinity
   J Atomic radius

7 Which of the following elements has the largest atomic radius?
   A Be
   B Mg
   C Ca
   D Sr

8 Which of the following elements has the smallest atomic radius?
   F Si
   G P
   H S
   J Cl

9 Which of the following elements has the largest atomic radius?
   A B
   B C
   C N
   D O

10 Which of the following elements has the lowest first ionization energy?
   F Na
   G Mg
   H Al
   J Si

11 Which of the following elements has the lowest first ionization energy?
   A Zr
   B Y
   C Sr
   D Rb

12 Which of the following elements has the greatest electronegativity?
   F F
   G Cl
   H Br
   J I

13 Which of the following elements has the greatest electronegativity?

A Al
B Si
C P
D S

14 Which of the following statements is true?

F The higher the ionization energy, the more reactive the metal.
G The lower the ionization energy, the more reactive the metal.
H The smaller the atomic radius, the more reactive the metal.
J The greater the electronegativity, the more reactive the metal.

15 Which of the following statements is true?

A The lower the ionization energy, the more reactive the nonmetal.
B The greater the atomic radius, the more reactive the nonmetal.
C The greater the electronegativity, the more reactive the nonmetal.
D The lower the electronegativity, the more reactive the nonmetal.

16 In an experiment, it is discovered that an element has a very low first ionization energy and a much larger second ionization energy. Of the following choices, the element was most likely—

F Ca
G Na
H Be
J P

17 All of the electrons in a particular group on the periodic table have the same

A electronegativities
B ionization energies
C number of valence electrons
D number of protons

Review for Lesson 3

18 Which of the following are most likely properties of a nonmetal?
- F High electronegativity and high ionization energy
- G High electronegativity and low ionization energy
- H Low electronegativity and high ionization energy
- J Low electronegativity and low ionization energy

19 An electron in an atom's outer shell is shielded from the nucleus by—
- A neutrons
- B protons
- C electrons in the inner shells of the atom
- D the electrons of other atoms

20 Which of the following statements is true?
- F Oxygen has a smaller electronegativity than carbon.
- G Oxygen has a lower first ionization energy than carbon.
- H Oxygen has a smaller atomic radius than carbon.
- J Oxygen has a smaller atomic weight than carbon.

21 Which of the following elements is an alkaline earth metal?
- A Rb
- B Ba
- C Ni
- D Se

22 Cobalt (Co) is—
   F an inert gas
   G an alkaline earth metal
   H a transition metal
   J a halogen

23 Which group contains the least reactive elements on the periodic table?
   A The noble gases
   B The halogens
   C The alkaline earth metals
   D The alkali metals

24 Which of the following best describes the element iodine?
   F Iodine is a halogen and a nonmetal.
   G Iodine is a halogen and a metal.
   H Iodine is a noble gas and a nonmetal.
   J Iodine is a noble gas and a metal.

25 Which of the following atoms has valence electrons in its *d* subshell?
   A Mg
   B Kr
   C Si
   D Ni

Review for Lesson 3

ANSWERS AND EXPLANATIONS

1. **C is correct.** Sulfur (S) is a nonmetal. Magnesium (Mg), iron (Fe), and lithium (Li) are metals. If you didn't remember, just look at the periodic table. Anything to the **right** of the heavy line is considered a nonmetal.

2. **G is correct.** Sodium (Na) is a metal. Phosphorous (P), neon (Ne), and oxygen (O) are nonmetals. Anything to the **left** of the heavy line is considered a metal.

3. **A is correct.** As (arsenic) is a metalloid because it is in the periodic table right next to the diagonal line that separates metals from nonmetals. Metalloids have properties of both metals and nonmetals.

4. **F is correct.** All of the elements listed are metals, so Rb is the most reactive because it is closest to the lower left-hand part of the periodic table.

5. **D is correct.** All of the elements listed are nonmetals, so Cl is the most reactive because it is closest to the upper right-hand part of the periodic table.

6. **G is correct.** Ionization energy tells how hard it is to remove an electron from an atom.

7. **D is correct.** Atomic radius increases as you move down a group on the periodic table.

8. **J is correct.** Atomic radius decreases as you move left to right across a period.

9. **A is correct.** Atomic radius decreases as you move left to right across a period, so B (boron) is the largest because it is farthest to the left.

10. **F is correct.** Ionization energy increases as you move left to right across a period, so Na (sodium) has the lowest ionization energy because it is farthest to the left.

11. **D is correct.** Ionization energy increases as you move left to right across a period, so Rb (rubidium) has the lowest ionization energy because it is farthest to the left.

12. **F is correct.** Electronegativity decreases as you move down a group. F (fluorine) is at the top of the group, so it has the greatest electronegativity. By the way, fluorine has the greatest electronegativity of all the elements.

13 D is correct. Electronegativity increases as you move across a period from left to right. S (sulfur) is farthest to the right, so it has the greatest electronegativity.

14 G is correct. Metals lose electrons in bonds, so the easier it is for a metal to give up an electron, the more reactive it will be. An element with a low ionization energy gives up an electron easily.

15 C is correct. Nonmetals gain electrons in bonds, so the greater a nonmetal's ability to take electrons away from another atom in a bond, the more reactive it will be. Electronegativity measures the ability of an atom to take another atom's electrons away.

16 G is correct. Na (sodium) has a very low first ionization energy because it is on the far left side of the periodic table. It has only one electron in its valence shell, so it is eager to get rid of it. But once sodium has lost an electron, it has a complete octet and is very stable. So the second ionization energy of sodium will be much larger than the first.

17 C is correct. Elements in the same group (that's a vertical column) have the same number of valence electrons.

18 F is correct. Nonmetals gain electrons in bonds, so they will have a large attraction for the electrons of other atoms (high electronegativity) and they will hang onto their own electrons tightly (high ionization energy).

19 C is correct. Shielding occurs when electrons in the inner shells of an atom act to repel electrons in the valence shell.

20 H is correct. Oxygen is farther to the right on the same period of the periodic table than carbon, so it has a smaller atomic radius. None of the other statements are true.

21 B is correct. Ba is in the second group from the left, so it is an alkaline earth metal.

22 H is correct. Cobalt is in the long, low section of the periodic table, so it is a transition metal.

23 A is correct. The noble gases, at the far right hand side of the table, have complete outer shells of electrons, so they do not need to form bonds with other atoms.

24 F is correct. Iodine (I) is in the halogen group and the nonmetal section of the periodic table.

25 D is correct. Ni (nickel) is a transition metal, so it has valence electrons in its *d* subshell.

LESSON 4
CHEMICAL COMPOUNDS

BONDING

Why do atoms bother to form bonds? Because they want a complete octet of valence electrons. As discussed earlier, this means having eight electrons in the outer shell. An element will gain, lose, or share electrons by forming bonds in order to fill their octet.

Atoms can come together and make **compounds** by forming two different kinds of bonds. In some cases, atoms will achieve an octet by completely giving up or gaining electrons. Bonds formed this way are called **ionic bonds**, and the compound is called an **ionic compound**. In other cases, atoms will share electrons in order to complete their octets. Bonds formed by sharing electrons are called **covalent bonds**, and the compound is called a **covalent compound**.

IONIC BONDS

When would atoms completely lose or gain an electron to form an ionic bond?

Ionic compounds form between atoms with very different electronegativities. In general, when a metal element bonds with a nonmetal element, an ionic bond is formed.

Think about the ionic compound sodium chloride, NaCl. Sodium is all the way to the left on the periodic table in Group I. It is a metal with fairly low electronegativity. Na atoms need to lose only one electron to form an octet. On the other hand, look at chlorine, all the way to the right of the periodic table. It is just one electron short of an octet. Chlorine is very electronegative and can easily pick up an extra electron and stabilize it because it has a high electron affinity. So NaCl sounds like a pretty compatible pair: Na would love to lose an electron and Cl would love to gain one. Once they exchange the electron, Na becomes a positively charged cation and Cl becomes a negatively charged anion.

When an atom gives up an electron, it becomes a positively charged ion, called a cation.

When an atom gains an electron, it becomes a negatively charged ion, called an anion.

Lesson 4: Chemical Compounds

In NaCl, the sodium cation and chloride anion are held together by electrostatic attraction. In ionic compounds, this attraction between the ions holds the atoms together. That's a fancy way of saying that the positive and negative ions stick together because opposites attract. In the picture below, a sodium atom has given up its single valence electron to a chlorine atom, which uses the electron to complete its outer shell. The two atoms are then held together by the attraction of the positive and negative charges on the ions.

$$[Na]^+[:\ddot{Cl}:]^-$$

You could probably guess that the larger the charges on the ions, the stronger the ionic bonds that will form. Therefore, an ionic bond between ions with +2 and –2 charges will be stronger than a bond between ions with +1 and –1 charges.

COVALENT BONDS
Covalent compounds form from atoms with similar electronegativities. In general, when two nonmetals bond together, a covalent bond is formed.

In a covalent bond, two atoms with similar electronegativities *share* electrons. Each atom counts the shared electrons as part of its valence shell, and in this way, both atoms can consider themselves to have complete outer shells.

In the picture below, two fluorine atoms, each of which needs one electron to complete its valence shell, form a covalent bond. Each atom donates an electron to the bond, which is considered part of the valence shell of both atoms.

$$:\!\ddot{F}\!\cdot\; +\; \cdot\!\ddot{F}\!: \;\Rightarrow\; :\!\ddot{F}\!:\!\ddot{F}\!:$$

It should make sense intuitively that a double bond is stronger than a single bond, and a triple bond is stronger than a double bond. Each bond is formed from sharing a pair of electrons, so a double bond has two pairs of shared electrons, and triple bond has three.

The first covalent bond formed between two atoms is called a sigma (σ) bond. All single bonds are sigma bonds. If additional bonds between the two atoms are formed, they are called pi (π) bonds. The second bond in a double bond is a pi bond, and the second and third bonds in a triple bond are pi bonds.

In the F_2 molecule shown on the previous page, the two fluorine atoms share the electrons equally because they have identical electronegativities, but that's not usually the case in molecules. Most covalent bonds have **polarity**. When atoms from different elements form a bond, the more electronegative atom will exert a stronger pull on the electrons in the bond. The pull is not enough to make the bond ionic, but it is enough to keep the electrons closer to one side of the molecule. This causes the molecule to have a **dipole**. That is, the side of the molecule where the electrons spend more time will be more negative and the side of the molecule where the electrons spend less time will be more positive. For example, in H_2O, the oxygen-hydrogen bonds have polarity. In which direction do you think the electrons are pulled? You know from the periodic trends that oxygen is the more electronegative element, and so the electrons are concentrated more toward the oxygen and farther from the hydrogens.

OXIDATION STATES

In both ionic and covalent bonds, each atom is assigned an **oxidation state** (or **oxidation number**). The oxidation state of an atom indicates the number of electrons that it gains or loses when it forms a bond. Even in a covalent bond where electrons are not completely lost or gained, each atom is given an oxidation number *as if* the electrons were not shared but completely given to the more electronegative atom as in an ionic bond. This is simply a way of keeping electrons counted as whole charges rather than partial charges. Before getting too wrapped up in rules, do an example to make this all clear. Consider water, H_2O. Although the bonded electrons are shared, for the sake of giving each atom an oxidation number, say that oxygen has a −2 charge and each hydrogen has a +1 charge. This is because oxygen is more electronegative so it attracts the two electrons from hydrogen toward itself. Each hydrogen is a little positive then. Notice that the charges must add up to zero, the total charge of the molecule. For an ionic compound, it's a straightforward process: The sodium atom in NaCl has a +1 charge and the chloride atom has a −1 charge. The oxidation number equals its charge. Again, the total charge adds up to zero because the total charge of NaCl is zero.

In summary, for an ionic bond, the atom that gains an electron gets a negative oxidation number and the atom that loses an electron gets a positive oxidation number. For a covalent bond, the atom with the greater electronegativity (that's the atom closest to the upper-right hand corner of the periodic table) gets a negative oxidation number and the other atom gets a positive number. Here are two important things you have to keep in mind when dealing with oxidation numbers:

The oxidation state of an atom that is not bonded to an atom of another element is zero.

That means an atom that is not bonded to any other atom, such as Ne, or an atom that is bonded to another atom of the same element, like the oxygen atoms in O_2 or the hydrogen atoms in H_2, have oxidation states of zero.

The oxidation numbers for all the atoms in a neutral molecule must add up to zero.

If you know that the oxidation state of oxygen in MgO is –2, then the oxidation state of Mg must be +2 to make the oxidation states add up to zero. If you know that the oxidation state of oxygen is –2 in H_2O, then you know that the oxidation state for each hydrogen atom must be +1, because there are two hydrogens in the molecule, and $1 + 1 - 2 = 0$.

For charged molecules, the oxidation state must add up to the charge.

For instance, in the hydroxide anion –OH, oxygen has an oxidation state of –2, and hydrogen has an oxidation state of +1. When you add it together, $-2 + 1 = -1$, you get the charge on the hydroxide molecule.

Most elements have different oxidation numbers that can vary depending on the molecule that they are a part of. The following chart shows some elements that consistently take the same oxidation numbers. There is no mystery to why these elements have consistent oxidation numbers. If you look at the periodic table, you'll see that by taking the oxidation numbers given below, each element achieves an octet. Keep the octet rule in mind, and you don't need to memorize these at all because you can predict from the periodic table the most stable oxidation state.

Element	Oxidation Number
Alkali metals (Li, Na, K...)	+1
Alkaline earths (Be, Mg, Ca...)	+2
Group 3A (B, Al, Ga...)	+3
Oxygen	–2
Halogens (F, Cl, Br...)	–1

As long as you know the oxidation numbers of the atoms in this chart (or can guess them from the periodic table), you can figure out the oxidation numbers of any other atoms in a molecule by using the fact that all of the oxidation numbers in a neutral molecule must add up to zero.

WHAT'S IN A NAME?—NAMING COMPOUNDS

CHEMICAL FORMULAS

A chemical formula shows the elements present in a compound using the symbols from the periodic table and number of atoms of each element present in the compound. For instance, the chemical formula for water, H_2O, tells you that a water molecule contains 2 hydrogen atoms and 1 oxygen atom. Similarly, a molecule of methane, CH_4, contains 1 carbon atom and 4 hydrogen atoms. There are four different methods of expressing a chemical formula. Each method gives slightly different information about the molecule.

The molecular formula tells the actual number of atoms of each element in a molecule.

If you see the molecular formula, C_4H_8, it means that this molecule contains 4 carbon atoms and 8 hydrogen atoms. The molecular formula is pretty straightforward: It just tells you exactly what a molecule is made of.

The empirical formula tells the lowest-whole-number ratio of elements in a molecule.

If an atom has the molecular formula C_4H_8, it will have the empirical formula CH_2, because there are twice as many hydrogen atoms as there are carbon atoms. The empirical formula gives less information than the molecular formula because many different molecules can have the same empirical formula. For example, molecules with the molecular formulas CH_2, C_2H_4, C_3H_6, C_4H_8, etc., will all have the same empirical formula, CH_2.

For some molecules, the empirical formula and the molecular formula are the same. This is true of molecules like H_2O, CO_2, and NH_3, whose molecular formulas also happen to be the lowest-whole-number ratios of the elements in the compound.

The structural formula shows the arrangements of the atoms and the bonds in a molecule.

The structural formula is a picture of the atom that shows the bonds between the atoms and gives you an idea of the shape of the molecule. The structural formulas for three molecules are shown below, along with their molecular formulas. The straight lines show bonds between atoms.

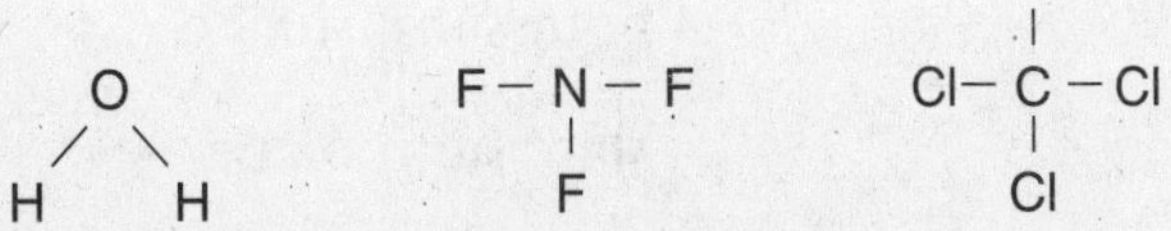

H_2O NF_3 CCl_4

Lesson 4: Chemical Compounds

The Lewis dot structure shows the arrangement of electrons in a molecule.

The Lewis dot structure gives the most information about a molecule. You draw a Lewis dot structure by taking a structural formula and adding all of the valence electrons for the different atoms in the molecule. You can see from the figures below how structural formulas are converted to Lewis dot structures. Each bond represents two electrons.

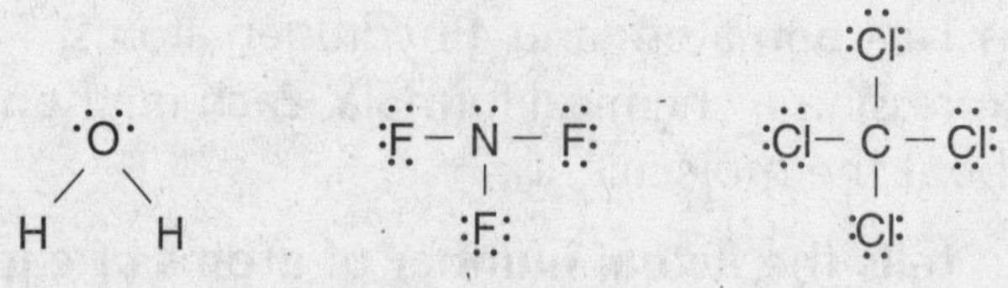

H_2O NF_3 CCl_4

Notice how each of the atoms in each molecule has a complete octet of electrons, except for the hydrogen atoms in H_2O, which only need two valence electrons. Remember that each bond attached to an atom counts as two valence electrons.

NAMING IONIC COMPOUNDS

Going from Formula to Name

An ionic compound always contains a metal and a nonmetal. If you know the formula of the compound, then getting the name is pretty straightforward. Here are the rules:

1. The metal ion (cation) always goes first and keeps its name.

2. The nonmetal ion (anion) goes second and changes its ending to *ide*.

3. You *don't* have to say how many of each ion are in the compound.

For example:

NaCl	sodium chloride
CaO	calcium oxide
MgF_2	magnesium fluoride
K_2S	potassium sulfide
Al_2O_3	aluminum oxide

Notice that the naming method is the same no matter how many of each ion are in the compound. That's because each of the metals and nonmetals above always forms the same ion and has the same oxidation state when it forms a bond.

There are times when the metals form different ions and have more than one oxidation state in an ionic bond. This occurs when the metal is a transition metal. In these cases, you will have to specify the oxidation state of the metal by using a Roman numeral in parentheses after the name of the metal.

For example, copper comes in two different ions, Cu^{1+} and Cu^{2+}. When copper bonds with chlorine, it can form two different compounds, $CuCl$ or $CuCl_2$. To distinguish between the two, you look at the oxidation state of copper in each compound. We can find the oxidation state of copper because we know that chlorine always gains one electron in a bond and has a -1 oxidation state. So in $CuCl$, copper has a $+1$ oxidation state, and in $CuCl_2$, copper has a $+2$ oxidation state.

$CuCl$	Cu^+Cl^-	copper(I) chloride
$CuCl_2$	$Cu^{2+}(Cl^-)_2$	copper(II) chloride

Here's another one

FeO	$Fe^{2+}O^{2-}$	iron(II) oxide
Fe_2O_3	$(Fe^{3+})_2(O^{2-})_3$	iron(III) oxide

Notice that the Roman number after iron does *not* tell you how many iron atoms are in the compound. Instead, the Roman number tells you iron's *oxidation state*.

Going from Name to Formula

If you know the name of the compound and you want to get the formula, you can use the crisscross method. Here's how it works:

1 Write the symbols of the metal and nonmetal in the compound.

2 Write the oxidation states or the ionic charge above the metal and nonmetal.

3 Use the crisscross method to get the formula.

Try it with calcium chloride.

1 $CaCl$

2 $Ca^{2+}Cl^{1-}$

3 Here's the crisscross:

$$Ca^{2+} \searrow Cl^{1-}$$
$$\phantom{Ca^{2+}}_{①} _{②}$$

So the formula for calcium chloride is $CaCl_2$.

Lesson 4: Chemical Compounds

Here's another, nickel(III) oxide. Remember, the Roman numeral tells you nickel's oxidation state.

1. NiO

2. $Ni^{3+} + O^{2-}$

3. Crisscross

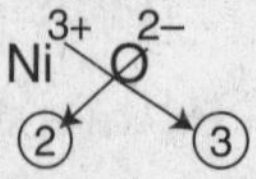

$$Ni^{3+} \underset{2}{\searrow} O^{2-} \underset{3}{\searrow}$$

So the formula for nickel(III) oxide is Ni_2O_3.

NAMING MOLECULAR COMPOUNDS

Going from Formula to Name

Molecular compounds are formed when nonmetals are joined together by covalent bonds. Here are the rules for getting the name of a compound if you know the formula:

1. The first element in the formula goes first in the name and it keeps its name.

2. The second element in the formula goes next and it changes its ending to *ide*.

3. The number of atoms of each element in the compound is indicated by a prefix.

You can omit the prefix *mono* if there is only one of the first element.

Prefix	Number of Atoms
Mono-	1
Di-	2
Tri-	3
Tetra-	4
Penta-	5
Hexa-	6
Hepta-	7
Octa-	8
Nona-	9
Deca-	10

For example:

SO$_2$ is sulfur dioxide. Notice that the prefix mono is not used for the first element.

SO$_3$ is sulfur trioxide.

P$_2$O$_5$ is diphosphorous pentoxide.

Cl$_2$O$_7$ is dichlorine heptoxide.

CO is carbon monoxide. Notice that the prefix mono is used for the second element.

Going from Name to Formula

You can find the formula of a molecular compound from its name. For example, the name nitrogen dioxide tells you that the compound contains 1 nitrogen atom and 2 oxygen atoms, so its formula is NO$_2$. Similarly, the name diphosphorous trioxide tells you that the compound contains 2 phosphorous atoms and 3 oxygen atoms, so the formula is P$_2$O$_3$.

NAMING POLYATOMIC IONS

Sometimes, atoms will join together to form an ion instead of a molecule. When that happens, the entire group of atoms will stick together and behave as a single ion, called a **polyatomic ion**. You should be familiar with the names and charges of the polyatomic ions listed below.

You need to memorize these or look them up because they are difficult to determine the charge.

Name	Formula and Charge
Nitrate	NO$_3^-$
Carbonate	CO$_3^{2-}$
Sulfate	SO$_4^{2-}$
Phosphate	PO$_4^{3-}$
Hydroxide	OH$^-$
Ammonium	NH$_4^+$

Lesson 4: Chemical Compounds

Naming Acids

An acid is a compound that will give up a hydrogen ion in solution. The rules for naming acids are a bit quirky, but you should be fine if you just make yourself familiar with the names and formulas of the most common acids, which are listed below.

Acid	Formula
Hydrochloric acid	HCl
Hydrobromic acid	HBr
Hydroiodic acid	HI
Nitric acid	HNO_3
Sulfuric acid	H_2SO_4
Carbonic acid	H_2CO_3
Phosphoric acid	H_3PO_4

REVIEW FOR
LESSON 4

1. Which of the following compounds contains an ionic bond?

   A CO
   B H_2O
   C NaCl
   D F_2

2. Which of the following compounds contains a covalent bond?

   F NO
   G MgO
   H KBr
   J CaS

3. Which of the following compounds contains a nonpolar covalent bond?

   A HCl
   B CaO
   C NH_3
   D O_2

4. Which of the following is true of a polar covalent bond?

   F It involves a positive ion (cation) and a negative ion (anion).
   G It involves equal sharing of electrons between two nonmetals.
   H It involves unequal sharing of electrons between two nonmetals.
   J It involves a metal and a nonmetal.

5. What is the oxidation state of magnesium in MgS?

   A +1
   B +2
   C −1
   D −2

6 Which of the elements below always takes an oxidation state of −1 when it forms a bond with another element?
   F Li
   G Be
   H O
   J F

7 What is the oxidation state of nitrogen in NO?
   A −1
   B −2
   C +1
   D +2

8 What is the oxidation state of nitrogen in NO_2?
   F +4
   G +3
   H +2
   J +1

9 What is the oxidation state of nitrogen in N_2O?
   A −1
   B −2
   C +1
   D +2

10 What is the oxidation state of nitrogen in Li_3N?
   F −1
   G −2
   H −3
   J −4

11 When an atom has an oxidation state of +2 in a compound, it means that the atom has—
   A gained 1 electron in a bond
   B gained 2 electrons in a bond
   C lost 1 electron in a bond
   D lost 2 electrons in a bond

12 What will be the oxidation state of an atom that has gained one electron in a bond?

F 0
G −1
H +1
J +2

13 What is the oxidation state of a single uncombined carbon atom?

A 0
B −1
C +1
D +2

14 What is the oxidation state of nitrogen in N_2?

F 0
G −1
H +1
J +2

15 What is the oxidation state of nitrogen in Al_2N_3?

A −1
B −2
C −3
D −4

16 Which of the following could be an empirical formula of a compound?

F C_2H_4
G C_3H_5
H C_3H_6
J C_4H_6

Review for Lesson 4

17 Which of the following is the structural formula of water?

   A HO

   B H_2O

   C H – O

   D H∖O∕H

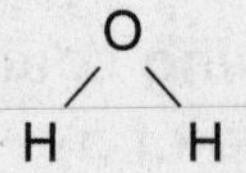

18 Which of the following is the Lewis dot structure of hydrochloric acid?

   F HCl

   G H – Cl

   H H – C̈l:

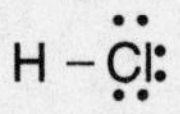

   J :Ḧ – C̈l:

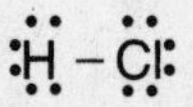

19 Which of the following is the name of the compound represented by KBr?

   A Potassium bromine
   B Potassium bromide
   C Bromine potassium
   D Bromide potassium

20 Which of the following is the name of the compound represented by $AlCl_3$?

   F Aluminum chloride
   G Aluminum chlorine
   H Trialuminum chlorine
   J Trichlorine aluminum

21 What is the formula for lithium bromide?

   A $LiBr_3$
   B $LiBr_2$
   C LiBr
   D Li_2Br

22. What is the formula for beryllium fluoride?
   F Be$_3$F
   G Be$_2$F
   H BeF
   J BeF$_2$

23. What is the formula for chromium(III) oxide?
   A CrO
   B Cr$_3$O
   C CrO$_3$
   D Cr$_2$O$_3$

24. What is the formula for chromium(II) oxide?
   F CrO
   G Cr$_2$O
   H CrO$_2$
   J Cr$_2$O$_3$

25. What is the name of the compound represented by N$_2$F$_4$?
   A Nitrogen fluorine
   B Nitrogen fluoride
   C Dinitrogen tetrafluoride
   D Dinitrogen fluoride

26. What is the name of the compound represented by the structural diagram shown below?

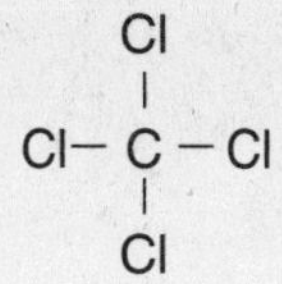

$$\begin{array}{c} \text{Cl} \\ | \\ \text{Cl}-\text{C}-\text{Cl} \\ | \\ \text{Cl} \end{array}$$

   F Carbon chlorine
   G Carbon chloride
   H Tetracarbon chloride
   J Carbon tetrachloride

27. What is the molecular formula of tetraphosphorous hexoxide?
   A P$_4$O$_6$
   B P$_2$O$_4$
   C P$_4$O
   D PO

Review for Lesson 4

28. What is the molecular formula of carbon disulfide?
   F C_2S_2
   G C_2S
   H CS_2
   J CS

29. What is the name of the compound represented by $NaNO_3$?
   A Sodium nitrate
   B Sodium nitrogen
   C Sodium nitride
   D Sodium nitrogen oxide

30. What is the name of the compound represented by HNO_3?
   F Hydrogen nitrogen oxide
   G Nitric acid
   H Nitrogen acid
   J Nitrogen oxygen acid

ANSWERS AND EXPLANATIONS

1. **C is correct.** NaCl is the only compound listed that has a metal bonded to a nonmetal. When a metal combines with a nonmetal, an ionic bond is formed.

2. **F is correct.** NO is the only compound listed that has a nonmetal bonded to a nonmetal. When a nonmetal combines with a nonmetal, a covalent bond is formed.

3. **D is correct.** A polar covalent bond is formed by two nonmetals with different electronegativities, so a nonpolar covalent bond will be formed by two nonmetals with equal electronegativities. All oxygen atoms have the same electronegativity, so the covalent bond that holds O_2 together is nonpolar.

4. **H is correct.** When nonmetals combine, they form covalent bonds. If the two nonmetals share their bonding electrons unequally, the covalent bond is polar.

5. **B is correct.** According to the periodic table, magnesium is in Group 2, so it always takes a +2 oxidation state when it forms a bond. If all you can remember is that Mg is a metal and that metals take positive oxidation states, you can still eliminate choices **C** and **D**.

6. **J is correct.** According to the periodic table, fluorine always takes a −1 oxidation state when it forms a bond with another element. Use POE to eliminate metals, which take positive oxidation states. That gets rid of choices **F** and **G**.

7. **D is correct.** You know from the periodic table that oxygen always takes a −2 oxidation state, so in order for the oxidation states in the NO molecule to add up to zero, the nitrogen atom must have a +2 oxidation state.

8. **F is correct.** You know from the periodic table that oxygen always takes a −2 oxidation state. There are two oxygen atoms in NO_2, so that adds up to −4. In order for the oxidation states of all the atoms in NO_2 to add up to zero, the nitrogen atom must have a +4 oxidation state.

9. **C is correct.** We know that oxygen always takes a −2 oxidation state. There are two nitrogen atoms in N_2O, so in order for the oxidation states of all the atoms to add up to zero, each of the nitrogen atoms must have an oxidation state of +1.

10. **H is correct.** According to the periodic table, lithium is in Group 1, so it always takes a +1 oxidation state. There are 3 lithium atoms in Li_3N, so that adds up to +3. In order for the oxidation states of all the atoms in Li_3N to add up to zero, the nitrogen atom must have a −3 oxidation state.

11 **D is correct.** Electrons are negatively charged, so if an atom loses electrons, it loses negative charge and ends up with a positive charge overall. If an atom has lost two electrons, it ends up with a +2 oxidation state. If you are guessing, you should eliminate choices **A** and **C**, because the answer is more likely to have a 2 than a 1 in it.

12 **G is correct.** Electrons are negatively charged, so if an atom gains electrons, it ends up with a negative charge. If an atom has gained one electron, it ends up with a −1 oxidation state. If you are guessing, you should eliminate choices **F** and **J**, because the answer probably will involve a 1.

13 **A is correct.** An atom that has not combined with another atom has not gained or lost electrons, so it has an oxidation state of zero.

14 **F is correct.** When two atoms of the same element combine to form a diatomic molecule (diatomic means two atoms), they share electrons equally. Because neither element can be thought of as gaining or losing electrons, the oxidation state of nitrogen in N_2 is zero. This is also true of all other compounds where two atoms of the same element combine to form diatomic molecules, such as O_2, H_2, and Cl_2. The atoms in all of these molecules have oxidation states of zero.

15 **B is correct.** Aluminum takes a +3 oxidation state in a bond. There are 2 aluminum atoms in Al_2N_3, so that adds up to +6. There are 3 nitrogen atoms in Al_2N_3, so in order for the oxidation states of all the atoms in Al_2N_3 to add up to zero, each of the nitrogen atoms must have a −2 oxidation state.

16 **G is correct.** The empirical formula gives the lowest whole number ratio of the elements in a compound. C_3H_5 is the only choice listed that cannot be reduced to lower whole number ratios.

17 **D is correct.** The molecular formula for water is H_2O. The structural formula shows the arrangement of the bonds in the molecule, so choice **D** is correct.

18 **H is correct.** The Lewis dot structure shows the bonds and arrangement of electrons in a molecule. Choice **H** is correct because it shows 2 electrons (the ones in the bond) connected to the hydrogen atom and 8 electrons connected to the chlorine atom. Choice **J** is wrong because hydrogen will never have 8 electrons around it.

19 **B is correct.** KBr is composed of a metal and a nonmetal, so it is an ionic compound. In an ionic compound, the metal is named first and the nonmetal goes second, with its ending changed to *ide*. KBr is called potassium bromide.

20 **F is correct.** $AlCl_3$ is composed of a metal and a nonmetal, so it is an ionic compound. In an ionic compound, the metal is named first and the nonmetal goes second, with its ending changed to *ide*. $AlCl_3$ is called aluminum chloride. Remember, aluminum has a +3 oxidation state, so you do not need numeric prefixes.

21 **C is correct.** You know that lithium gives up 1 electron in a bond and takes a +1 oxidation state and that bromine gains 1 electron and takes a −1 oxidation state. Lithium bromide is an ionic compound, so find the formula by using the crisscross method.

$$Li^{+1} \searrow\!\!\!\!\!\nearrow Br^{-1}$$
$$①①$$

You don't have to write 1s in the formula, so the formula for lithium bromide is LiBr.

22 **J is correct.** You know that beryllium gives up 2 electrons in a bond and takes a +2 oxidation state and that fluorine gains 1 electron and takes a −1 oxidation state. Beryllium fluoride is an ionic compound, so find the formula by using the crisscross method.

$$Be^{+2} \searrow\!\!\!\!\!\nearrow F^{-1}$$
$$①②$$

You don't have to write 1s in the formula, so the formula for beryllium fluoride is BeF_2.

23 **D is correct.** Chromium is a transition metal, so the Roman numeral tells us that its oxidation state is +3. Oxygen gains 2 electrons in a bond and takes an oxidation state of −2. Chromium(III) oxide is an ionic compound, so find the formula using the crisscross method.

$$Cr^{+3} \searrow\!\!\!\!\!\nearrow O^{-2}$$
$$②③$$

The formula for chromium(III) oxide is Cr_2O_3.

24 **F** is **correct**. Chromium is a transition metal, so the Roman numeral tells us that its oxidation state is +2. Oxygen gains 2 electrons in a bond and takes an oxidation state of −2. Chromium(II) oxide is an ionic compound, so find the formula using the crisscross method.

Cr_2O_2 can be reduced to lowest whole number ratios of Cr_1O_1. The formula for chromium(II) oxide is CrO.

25 **C** is **correct**. Nitrogen and fluorine are nonmetals, so they form a molecular compound when they bond together. For molecular compounds, use prefixes to tell how many atoms of each element are in the compound. Because there are 2 nitrogen atoms and 4 fluorine atoms in the compound, the name is dinitrogen tetrafluoride.

26 **J** is **correct**. The diagram shows 1 carbon atom bonded to 4 chlorine atoms. Carbon and chlorine are nonmetals, so they form a molecular compound when they bond together. For molecular compounds, use prefixes to tell how many atoms of each element are in the compound. Because there is 1 carbon atom and 4 chlorine atoms in the compound, the name is carbon tetrachloride.

27 **A** is **correct**. You can get the formula of a molecular compound directly from its name. The prefixes in tetraphosphorous hexoxide tell you that the compound contains 4 phosphorous atoms and 6 oxygen atoms, so the formula is P_4O_6.

28 **H** is **correct**. You can get the formula of a molecular compound directly from its name. The prefixes in carbon disulfide tell you that the compound contains 1 carbon atom and 2 sulfur atoms, so the formula is CS_2.

29 **A** is **correct**. The compound contains sodium (Na) and the nitrate polyatomic ion (NO_3^-), so the name is just sodium nitrate.

30 **G** is **correct**. You should recognize HNO_3 as one of the acids that you should be familiar with. HNO_3 is nitric acid.

LESSON 5
CHEMICAL REACTIONS

CHEMICAL AND PHYSICAL CHANGES

When a substance undergoes a **chemical change**, a chemical reaction takes place and the substance is converted into a different substance. In a chemical change, the chemical formulas of the products will be different from those of the reactants. Some examples of chemical changes are burning wood, hard-boiling an egg, and rusting iron. It's very difficult to undo most chemical changes. You've never seen anyone unboil an egg or retrieve a piece of wood from smoke and ash.

A **physical change** takes place without a chemical reaction. In a physical change, the products have the same chemical formulas as the reactants. Phase changes are physical changes. Melting, freezing, boiling, and condensing are physical changes. It's usually not difficult to undo a physical change. You can melt an ice cube and then put the water back in the freezer to form ice again.

BALANCING ACT—CHEMICAL EQUATIONS

One of the basic rules of chemistry is that matter is never created or destroyed in a chemical reaction. That means that if there are 3 carbon atoms and 4 hydrogen atoms in the reactants, then there must be 3 carbon atoms and 4 hydrogen atoms in the products. In other words, all chemical reactions must be balanced. When you balance an equation, you are making sure that you have the same numbers of each atom in the reactants and the products.

Normally, balancing a chemical equation is a trial and error process. You start with the most complicated looking compound in the equation and work from there. There is, however, a technique that you might want to try if you see a balancing equation question. The technique is called *backsolving*.

Backsolving works like this: On any question that involves balancing equations, one of the answer choices will have the correct coefficient for one of the compounds in the reaction. So instead of starting blind in the trial and error process, you can insert the answer choices one by one to see which one works. You probably won't have to try all four. If you start with one of the middle choices and the number doesn't work, it might be obviously too small or large, you may be able to eliminate other choices without trying them.

Example 1 $\quad$ ___NH$_3$ + ___O$_2$ → ___N$_2$ + ___H$_2$O

If the equation above were balanced with lowest-whole-number coefficients, the coefficient for NH$_3$ would be

A   1
B   2
C   3
D   4

Start at choice C because it's in the middle. If there are 3 NH$_3$'s, then there can't be a whole number coefficient for N$_2$. Choice **C** is wrong, and so is the other odd number answer, choice **A**.

Try choice **D**. If there are 4 NH$_3$'s, then there must be 2 N$_2$'s and 6 H$_2$O's. If there are 6 H$_2$O's, then there must be 3 O$_2$'s, and the equation is balanced with lowest-whole-number coefficients. The correct answer choice is **D**.

$$4\,NH_3 + 3\,O_2 \rightarrow 2\,N_2 + 6\,H_2O$$

Backsolving is more efficient than the methods you're used to because the correct answer choice is there. All you have to do is find it. Try each answer choice and use process of elimination to get rid of wrong answers.

Example 2 $\quad$ ___Ag + ___H$_2$S → ___Ag$_2$S + ___H$_2$

When this equation is balanced, what is the coefficient of Ag?

F   1
G   2
H   3
J   4

Start with choice **H**. If there are 3 Ag's, then there can't be a whole number coefficient for Ag$_2$S, so choice **H** is wrong, and so is the other odd number answer, choice **F**.

Try choice **G**. If there are 2 Ag's, then there must be 1 Ag$_2$S. If there is 1 Ag$_2$S, then there must be 1 H$_2$S. If there is 1 H$_2$S, then there must be 1 H$_2$, and the equation is balanced with lowest-whole-number coefficients. Answer choice **G** is correct.

$$2\,Ag + 1\,H_2S \rightarrow 1\,Ag_2S + 1\,H_2$$

Some questions can't be worked out by backsolving.

Example 3 | ___C_3H_4 + ___O_2 → ___CO_2 + ___H_2O

When the equation above is balanced, what is the sum of the coefficients?
- A 4
- B 5
- C 8
- D 10

In this question, you are asked for the sum of the coefficients, so you can't backsolve. Instead, you should use trial and error by choosing a coefficient for one of the compounds. The best way to do this is to choose 1 as the coefficient for the most complicated compound and see if it works. If it doesn't work, then try 2, and so on.

If there is 1 C_3H_4, then there are 3 CO_2's and 2 H_2O's. Now count the O's on the right-hand side and get 8. That means that there are 4 O_2's. That works and the equation is balanced with lowest-whole-number coefficients.

$$1 \, C_3H_4 + 4 \, O_2 \rightarrow 3 \, CO_2 + 2 \, H_2O$$

Now you can add up the coefficients, 1 + 4 + 3 + 2 = 10, choice **D**.

CHEMICAL REACTIONS

There are too many different kinds of chemical reactions for you to be able to keep track of, but you should be able to recognize a handful of the basic reactions that occur. Most of the reactions you've heard of will fall into the groups listed below.

SYNTHESIS

In a synthesis or combination reaction, *two or more reactants come together to form a single product*. The reaction below shows how hydrogen and oxygen come together to form water. Don't worry about the coefficients, the important thing here is that two kinds of reactants come together to form one kind of product.

$$2 \, H_2 + O_2 \rightarrow 2 \, H_2O$$

The reaction above is a typical synthesis reaction, with two reactants combining to form one product. Here are a few more:

$$3 \, H_2 + N_2 \rightarrow 2 \, NH_3$$
$$C + O_2 \rightarrow CO_2$$
$$2 \, Ca + O_2 \rightarrow 2 \, CaO$$

To recognize a synthesis reaction, just look for a reaction with only one kind of product.

DECOMPOSITION

A decomposition reaction is the opposite of a synthesis reaction. *In a decomposition reaction, a single reactant will break apart and form two or more products.* You usually have to add heat to the reactant to get a decomposition reaction to happen. The reaction below shows the decomposition of hydrochloric acid into hydrogen and chlorine gas.

$$2\ HCl \rightarrow H_2 + Cl_2$$

Some chemical reactions are reversible, so some decomposition reactions are just the synthesis reactions shown above written in reverse.

$$2\ H_2O \rightarrow 2\ H_2 + O_2$$
$$2\ NH_3 \rightarrow 3\ H_2 + N_2$$

Here are a few more:

$$2\ KClO_3 \rightarrow 2\ KCl + 3\ O_2$$
$$H_2O_2 \rightarrow 2\ H_2 + O_2$$

To spot a decomposition reaction, just look for the reaction with only one kind of reactant.

SINGLE REPLACEMENT

In a single replacement reaction, an atom in a compound is replaced by an atom of a different element. In the reaction below, a potassium ion replaces the sodium ion in sodium bromide.

$$K^+ + NaBr \rightarrow Na^+ + KBr$$

Notice that potassium and sodium cations have the same charge. They can be replaced for each other.

Here's another one:

$$Br^- + CH_3I \rightarrow CH_3Br + I^-$$

In this one, the Br^- ion replaces the I to give an I^- anion. This is simply a replacement of one anion with another.

$$Cu^{2+} + 2\ AgNO_3 \rightarrow 2\ Ag^+ + Cu(NO_3)_2$$

This one looks a little more complicated, but the same thing is happening. The Cu^{2+} ion replaces the Ag^+ ion in the compound. Because the silver and copper cations have different charges, they cannot simply replace each other. Think about what is formed when Cu^{2+} replaces Ag^+. The compound is $CuNO_3^-$. An ion is formed. To balance charges, both nitrate ions become bonded to Cu^{2+} to form $Cu(NO_3)_2$. You are left with two silver ions.

Here are a couple more:

$Mg^{2+} + 2\ NaCl \rightarrow 2\ Na^+ + MgCl_2$

$NH_4^+ + H_2CO_3 \rightarrow NH_4HCO_3 + H^+$

The last one is also a single replacement reaction, although it looks a little different because it involves the polyatomic cation, ammonium. Notice how NH_4^+ replaces one of the protons, H^+.

DOUBLE REPLACEMENT

In a double replacement reaction, atoms in two reactant compounds trade places, kind of like trading partners at a dance. Here's a typical one.

$AgNO_3 + NaCl \rightarrow AgCl + NaNO_3$

Many double replacement reactions are **precipitation reactions**. In a precipitation reaction, two solutions are mixed, and as one of the products is formed, it precipitates out from the solution because it is insoluble. In the case above, the solid is AgCl. Notice how the charges of the anions and cations involved in the replacement are balanced. For example, the silver and sodium cations are both +1 and the nitrate and chloride anions are both –1. The anions and cations can easily exchange.

Here are a few more:

$BaCl_2 + K_2SO_4 \rightarrow BaSO_4 + 2\ KCl$

In this example, the cations and anions have different charges. The barium cation has a charge of +2 and the potassium cation has a charge of +1. The chloride anion has a charge of –1 and the sulfate anion has a charge of –2. When the ions replace each other, $BaSO_4$ and KCl are formed. There is an extra potassium ion and an extra chloride ion left over. They form another KCl molecule.

$Pb(NO_3)_2 + CaSO_4 \rightarrow Ca(NO_3)_2 + PbSO_4$

In these examples, the salts that precipitate out of the solution are $BaSO_4$ and $PbSO_4$.

NEUTRALIZATION

In a neutralization reaction, *an acid and a base react to form water and a salt as products*. In the reaction below, hydrochloric acid (HCl) reacts with sodium hydroxide (NaOH, a base) to form water (H_2O) and sodium chloride (NaCl, a salt).

$HCl + NaOH \rightarrow H_2O + NaCl$

A neutralization reaction is really just a special kind of double replacement reaction. In the case above, the H atom in HCl and the Na atom in NaOH trade places. Here are a few more.

$HNO_3 + KOH \rightarrow H_2O + KNO_3$

Lesson 5: Chemical Reactions

$$2\ HBr + Ca(OH)_2 \rightarrow 2\ H_2O + CaBr_2$$

Notice that *water is always a product of a neutralization reaction*.

COMBUSTION

In combustion, better known as burning, *a compound reacts with oxygen to produce water and carbon dioxide as products*. The reaction below shows the combustion of methane (CH_4) gas, which is what happens when you turn on a gas stove.

$$CH_4 + 2\ O_2 \rightarrow 2\ H_2O + CO_2$$

Combustion reactions release a lot of energy, which means that they are **exothermic** (reactions that take in energy are **endothermic**). Here are a few more combustion reactions.

$$C_2H_4 + 3\ O_2 \rightarrow 2\ H_2O + 2\ CO_2$$
$$C_3H_8 + 5\ O_2 \rightarrow 4\ H_2O + 3\ CO_2$$

You can see that most combustion reactions look pretty similar to each other.

OXIDATION-REDUCTION

In an oxidation-reduction reaction, certain atoms involved in the reaction gain or lose electrons. When an atom gains or loses electrons, its oxidation number changes. If an atom gains electrons, it becomes more negative and its oxidation number goes down. When an atom's oxidation number goes down, we say that it has been **reduced**. When an atom loses electrons, it becomes more positively charged and its oxidation number goes up. When an atom's oxidation number goes up, we say that it has been **oxidized**. Make a note of that.

When an atom is *reduced*, its oxidation number goes *down*.

When an atom is *oxidized*, its oxidation number goes *up*.

Oxidation and reduction always happen together in a reaction. So in any oxidation-reduction (redox for short) reaction, there will always be both an oxidation and a reduction taking place. Look at the reaction types you've seen already and identify which ones are also oxidation-reduction reactions.

A synthesis reaction is also an oxidation-reduction reaction.

$$2\ \overset{0}{H_2} + \overset{0}{O_2} \rightarrow 2\overset{21+\ 2-}{H_2O}$$

In the synthesis reaction above, H and O have oxidation states of zero at the beginning of the reaction. That's because H_2 and O_2 are elements in their uncombined state, and the oxidation state of any uncombined element is zero. When H and O combine to produce H_2O, the oxidation state of H goes from 0 to +1, so H is oxidized in the reaction. The oxidation state of O goes from 0 to –2, so O is reduced in the reaction.

- A decomposition reaction is also an oxidation-reduction reaction.

$$\overset{1+\ 1-}{2\ HCl} \rightarrow \overset{0\ \ 0}{H_2\ Cl_2}$$

In the decomposition reaction above, the oxidation state of H goes from +1 to 0, so H is reduced. The oxidation state of Cl goes from –1 to 0, so Cl is oxidized.

- A single replacement reaction can also be an oxidation-reduction reaction.

$$\overset{0}{K} + \overset{1+}{Na} + \overset{1-}{Br} \rightarrow \overset{0}{Na} + \overset{1+\ 1-}{K\,Br}$$

In the single replacement reaction shown above, the oxidation state of K goes from 0 to +1, so K is oxidized. The oxidation state of Na goes from +1 to 0, so Na is reduced. Sometimes, redox reactions are written showing only the atoms that are oxidized and reduced, so the reaction above could also be written this way:

$$\overset{0}{K} + \overset{1+}{Na} \rightarrow \overset{0}{Na} + \overset{1+}{K}$$

In any oxidation-reduction reaction, the number of electrons lost by one atom must be the same as the number of electrons gained by the other atom. Sometimes, this fact is used when balancing oxidation reactions.

- A double replacement reaction is NOT an oxidation-reduction reaction.

$$\overset{1+\ \ 1-}{Ag(NO_3)} + \overset{1+\ 1-}{Na\,Cl} \rightarrow \overset{1+\ 1-}{Ag\,Cl} + \overset{1+\ \ 1-}{Na(NO_3)}$$

In the double replacement reaction shown above, none of the oxidation numbers are changed, so no electrons are gained or lost in the reaction and none of the atoms are oxidized or reduced.

- A neutralization reaction is NOT an oxidation-reduction reaction.

$$\overset{1+\ 1-}{HCl} + \overset{1+\ \ 1-}{Na(OH)} \rightarrow \overset{1+\ \ 1-}{H(OH)} + \overset{1+\ 1-}{Na\,Cl}$$

Because a neutralization reaction is a kind of double replacement reaction, it makes sense that it is also not an oxidation-reduction reaction. Notice that water in the products is written as HOH instead of as H_2O in order to make it clearer that none of the oxidation states are changing.

- A combustion reaction is an oxidation-reduction reaction.

$$\overset{4-\ 4(1+)}{C\,H_4} + \overset{0}{2\ O_2} \rightarrow \overset{2(1+)\ 2-}{2\,H_2\,O} + \overset{4+\ 2(2-)}{C\,O_2}$$

This one is a little bit strange to look at because CH_4 is written with the negative oxidation state in front (normally the positive oxidation state is placed first). In this combustion reaction, the oxidation state of C changes from –4 to +4, so C is oxidized. The oxidation state of O goes from 0 to –2, so O is reduced.

Lesson 5: Chemical Reactions

NUCLEAR REACTIONS

Nuclear reactions are different from chemical reactions. In a chemical reaction, electrons are exchanged among different atoms, but the nuclei of the atoms are unaffected. In a nuclear reaction, the nuclei of atoms are changed. There are two main types of nuclear reactions, fission and fusion.

FISSION

In a fission reaction, a larger nucleus is broken up into smaller nuclei. An example of a fission reaction is shown below.

$$_0^1n + {}_{92}^{235}U \rightarrow {}_{56}^{142}Ba + {}_{36}^{91}Kr + 3\,_0^1n$$

In the reaction above, a neutron ($_0^1n$) collides with a uranium atom and breaks it into two other atoms and 3 nuclei. All nuclear energy produced in nuclear reactors today is made by the process of fission.

FUSION

In a fusion reaction, *two smaller nuclei come together to form a larger nucleus.* An example of a fusion reaction is shown below.

$$_1^2H + {}_1^3H \rightarrow {}_2^4He + {}_0^1n$$

In the reaction above, two isotopes of hydrogen combine to form helium and leave a spare neutron. The sun produces energy through fusion. Nuclear fusion requires extremely high temperatures, so it has not been practical as a way to produce energy on earth.

BALANCING NUCLEAR REACTIONS

To balance a nuclear reaction, you need to make sure that the mass numbers and atomic numbers are balanced. **Remember**, the mass numbers are the superscripts for the elements and the atomic numbers are the subscripts. For instance, look again at the example of a fission reaction.

$$_0^1n + {}_{92}^{235}U \rightarrow {}_{56}^{142}Ba + {}_{36}^{91}Kr + 3\,_0^1n$$

The sum of the mass numbers on the left side of the reaction is equal to the sum of the mass numbers on the right side of the reaction.

$$1 + 235 = 142 + 91 + 3(1) = 236$$

The sums of the atomic numbers are also equal.

$$0 + 92 = 56 + 36 + 3(0) = 92$$

The same is true for the fusion reaction.

$$^{2}_{1}H + ^{3}_{1}H \rightarrow ^{4}_{2}He + ^{1}_{0}n$$

Mass numbers: $2 + 3 = 4 + 1 = 5$
Atomic numbers: $1 + 1 = 2 + 0 = 2$

RADIOACTIVE DECAY

You may know **radioactive elements** such as uranium-238 (^{238}U) and carbon-14 (^{14}C). These special isotopes have unstable nuclei, that is, certain combinations of protons and neutrons just don't like to be together. To find a combination that is stable, the nuclei of these isotopes will actually emit particles of protons and neutrons, and even electrons, in order to stabilize their nuclei. This process of changing the proton/neutron ratio in the nucleus is called **radioactive decay**. There are two common types of radioactive decay: Emission of alpha particles and emission of beta particles.

ALPHA DECAY

An alpha particle is two neutrons and two protons. What is the charge on an alpha particle? Two protons give a charge of +2, and the neutrons don't add any charge, so the charge on the alpha particle is +2. An example of alpha decay is radium-226 to radon-222:

$$^{226}_{88}Ra \rightarrow ^{222}_{86}Rn + ^{4}_{2}a$$

During this decay, an alpha particle is given off. The alpha particle has two protons, so it decreases the atomic number of radium by two. The atomic number is now $88 - 2 = 86$, the atomic number for radon. Remember when the atomic number changes, the element changes as well. The alpha particle has an atomic weight of 4, so the mass number changes from 226 to 222. The chemical equation is now balanced in terms of atomic weight and atomic number.

When a nucleus undergoes alpha decay, the atomic number decreases by two and the atomic weight decreases by four.

Beta Decay

A beta particle is an electron. When an atom undergoes beta decay, an electron is ejected from the nucleus and a neutron becomes a proton. Because a neutron becomes a proton, the atomic number of the element increases by 1. A beta particle, or an electron, has very little mass, so the mass number does not change.

The example below shows the decay of carbon-14 to nitrogen-14. Notice that only the atomic number changes; the mass number stays the same. The sum of the atomic numbers and the sum of the atomic masses on each side of the reaction are equal.

$$^{14}_{6}C \rightarrow {}^{14}_{7}N + {}^{0}_{-1}\beta$$

You have to balance the atomic numbers and atomic weights and identify what new element you've made.

When a nucleus undergoes beta decay, the atomic number increases by one and the atomic weight remains the same.

REVIEW FOR LESSON 5

1

___Fe + ___O_2 → ___Fe_2O_3

When the equation above is properly balanced, what is the coefficient for Fe_2O_3?

A 1
B 2
C 3
D 4

2

___$KClO_3$ → ___KCl + ___O_2

When the equation above is properly balanced, what is the coefficient for $KClO_3$?

F 4
G 3
H 2
J 1

3

___C_3H_8 + ___O_2 → ___CO_2 + ___H_2O

When the equation above is properly balanced, what is the coefficient for CO_2?

A 3
B 5
C 6
D 9

4

___F_2 + ___$SnCl_2$ → ___SnF_2 + ___Cl_2

When the equation above is properly balanced, what is the sum of the coefficients?

F 10
G 8
H 6
J 4

5

___N_2O_5 → ___NO_2 + ___O_2

When the equation above is properly balanced, what is the coefficient for N_2O_5?

A 1
B 2
C 3
D 4

Review for Lesson 5

6.

$$2\,Al + 3\,Cl_2 \rightarrow 2\,AlCl_3$$

Which of the following best describes the reaction shown above?

F Decomposition
G Synthesis
H Single replacement
J Double replacement

7.

$$Ca(NO_3)_2 + Na_2CO_3 \rightarrow CaCO_3 + 2\,NaNO_3$$

Which of the following best describes the reaction shown above?

A Combustion
B Oxidation-reduction
C Single replacement
D Double replacement

8. Which of the following is a decomposition reaction?

F $HBr + KOH \rightarrow H_2O + KBr$
G $Zn + CuSO_4 \rightarrow ZnSO_4 + Cu$
H $2\,PbO \rightarrow 2\,Pb + O_2$
J $2\,C_2H_2 + 3\,O_2 \rightarrow 2\,H_2O + 2\,CO_2$

9. All of the following statements about oxidation-reduction reactions are true except—

A a compound must get reduced
B a compound must get oxidized
C electrons must be exchanged
D oxygen is always involved

10. Which of the following is a neutralization reaction?

F $H_2SO_4 + Mg(OH)_2 \rightarrow 2\,H_2O + MgSO_4$
G $C_3H_4 + 4\,O_2 \rightarrow 2\,H_2O + 3\,CO_2$
H $Cl_2 + CaBr_2 \rightarrow CaCl_2 + Br_2$
J $Fe + 2\,AgNO_3 \rightarrow Fe(NO_3)_2 + 2\,Ag$

ANSWERS AND EXPLANATIONS

1. **B is correct.** Backsolve. If there are 2 Fe_2O_3's, then there must be 4 Fe's and 3 O_2's. That balances the equation with lowest whole number coefficients.

$$4\ Fe + 3\ O_2 \rightarrow 2\ Fe_2O_3$$

If you answered choice **D**, you found whole number coefficients but not the *lowest* whole number coefficients. You need lowest whole number coefficients to balance an equation properly.

2. **H is correct.** Backsolve. If there are 2 $KClO_3$'s, then there must be 2 KCl's and 3 O_2's. That balances the equation with lowest whole number coefficients.

$$2\ KClO_3 \rightarrow 2\ KCl + 3\ O_2$$

If you chose choice **F**, you found whole number coefficients but not the *lowest* whole number coefficients. You need lowest whole number coefficients to balance an equation properly.

3. **A is correct.** Backsolve. If there are 3 CO_2's, then there must be 1 C_3H_8. If there is 1 C_3H_8, then there must be 4 H_2O's. Now you can count the O's on the right-hand side and get a total of 10, so there must be 5 O_2's. That balances the equation with lowest whole number coefficients.

$$1\ C_3H_8 + 5\ O_2 \rightarrow 3\ CO_2 + 4\ H_2O$$

If you answered choice **C** or **D**, you found whole number coefficients but not the *lowest* whole number coefficients. You need lowest whole number coefficients to balance an equation properly.

4. **J is correct.** You can't backsolve on this one. If you try 1 as the coefficient for $SnCl_2$, you find that there must be 1 SnF_2, 1 Cl_2, and 1 F_2, so the equation is balanced and all of the coefficients are equal to 1.

$$1\ F_2 + 1\ SnCl_2 \rightarrow 1\ SnF_2 + 1\ Cl_2$$

So the sum of the coefficients is 4.

5. **B is correct.** Backsolve. If there are 2 N_2O_5's, then there must be 4 NO_2's.

Review for Lesson 5

Now you can count the O's on the both sides of the equation. There are 10 O's on the left and 8 on the right so far, so if you put 1 O_2 on the right, the equation is balanced.

$$2\ N_2O_5 \rightarrow 4\ NO_2 + 1\ O_2$$

If you answered choice **D**, you found whole number coefficients but not the *lowest* whole number coefficients. You need lowest whole number coefficients to balance an equation properly.

6   **G is correct.** In the reaction, two reactants, Al and Cl, combine to form a single product, $AlCl_3$, so the reaction is a synthesis reaction.

7   **D is correct.** In the reaction, Ca and Na change partners with NO_3 and CO_3, so the reaction is a double replacement reaction. Remember that double replacement reactions are not oxidation-reduction reactions. If you're not sure, you can eliminate choice **A**, combustion, because it doesn't look like anything is burning in this reaction.

8   **H is correct.** In choice **H**, a single reactant, PbO, breaks up into two products, Pb and O_2, so it is a decomposition reaction. Choice **F** is a neutralization reaction. Choice **G** is a single replacement reaction. Choice **J** is a combustion reaction.

9   **D is correct.** In an oxidation-reduction reaction, one atom always gives up electrons or gets oxidized and another atom always gains electrons or gets reduced. Electrons are exchanges, so you can eliminate answer choices **A**, **B**, and **C**. Although the term oxidized has the same root as oxygen, oxidation-reduction reactions do not always involve oxygen.

10 **F is correct.** In choice **F**, an acid, H_2SO_4, combines with a base, $Mg(OH)_2$, to form water, H_2O, and a salt, $MgSO_4$. That's a neutralization reaction. Choice **G** is a combustion reaction. Choices **H** and **J** are both single replacement reactions.

LESSON 6
CHEMISTRY COMPUTATION

UNITS
Chemists use the **International System (SI)**, so all of the measurements that you see will be in metric units. All the units in the metric system are in multiples of ten. Once you know the prefixes and suffixes, it's simple to convert between units. This makes the math much easier. However, you may be unfamiliar with the units on an everyday basis. That is, you know what it means to jump three feet or what 95 pounds weighs, but how far is a 10-km race or what does 300 grams weigh? This lesson introduces you to the basics of the metric system and gives you familiar examples of typical units.

BASIC UNITS
Length is measured in **meters**. A meter is about the same size as a yard. Other typical measurement unit sizes are centimeters and millimeters.

Weight is measured in **grams**. A gram is the weight of a raisin or a paper clip. Weight is often expressed in kilograms.

Volume is measured in **liters**. A liter is about the same size as a quart. You'll also see volume measured in milliliters.

Water serves to connect these units in a couple of different ways.

> 1 cubic centimeter of water has a volume of 1 milliliter.
>
> 1 milliliter of water weighs 1 gram.
>
> 1 liter of water weighs 1 kilogram.

PREFIXES
You should know the prefixes that are added to the basic units.

Prefix	Symbol	Meaning	Example
Giga-	G	10^9	There are 10^9 m in 1 Gm.
Mega-	M	10^6	There are 10^6 m in 1 Mm.
Kilo-	k	10^3	There are 1,000 m in 1 km.
Deci-	d	0.1	There are 10 dm in 1 m.
Centi-	c	0.01 or 10^{-2}	There are 100 cm in 1 m.
Milli-	m	0.001 or 10^{-3}	There are 10^3 mm in 1 m.
Micro-	μ	10^{-6}	There are 10^6 μm in 1 m.
Nano-	n	10^{-9}	There are 10^9 nm in 1 m.

CONVERTING

Dimensional analysis is a big term for a simple idea: converting units. You may remember the rule from algebra that when you multiply anything by 1, you don't change the value at all. Dimensional analysis boils down to multiplying by 1 to convert one unit to another. Here's how it's done:

You want to convert 250 centimeters to meters.

First, use an expression that tells how many centimeters there are in 1 meter.

$$100 \text{ cm} = 1 \text{ m}$$

Now, write it as a fraction equal to 1. When you make the fraction, always put the unit you want to get rid of on the bottom. In this case, you are converting from centimeters to meters, so you want to get rid of centimeters.

$$\frac{1 \text{ m}}{100 \text{ cm}}$$

Finally, you multiply the value you want to convert by the fraction you have created.

$$(250 \text{ cm})\left(\frac{1 \text{ m}}{100 \text{ cm}}\right) = \frac{250}{100} \text{ m} = 2.5 \text{ m}$$

The units you don't want will cancel and leave you with the units you want.

Here's another one. Convert 1.2 liters to milliliters.

First, find an expression that equates liters and milliliters.

$$1 \text{ L} = 1,000 \text{ mL}$$

Next, set it up as a fraction with liters on the bottom because that's the unit you want to get rid of.

$$\frac{1,000 \text{ ml}}{1 \text{ L}}$$

Now, multiply the measurement by the fraction.

$$(1.2 \text{ L})\left(\frac{1,000 \text{ mL}}{1 \text{ L}}\right) = (1.2)(1,000) \text{ mL} = 1,200 \text{ mL}$$

DENSITY

Density is a measure of mass per volume. For example, which has greater density, water or air? To answer this without math, think of measuring water and air in containers of the same size, say, a one-gallon jug. The volumes are equal, so the jug with the greater mass has greater density. Water is denser than air. The formula for density is below.

$$\text{density} = \frac{\text{mass}}{\text{volume}}$$

Density is usually measured in units of grams/cm^3 or kg/m^3.

SIGNIFICANT DIGITS 101

In chemistry, all numbers must be presented with the proper number of significant digits. The number of significant digits in a measurement indicates the accuracy of the measurement. For instance, if you see a measurement of 1 meter, it means that the measurement was taken to the nearest meter, but if you see a measurement of 1.00 meter, it means that the measurement is accurate to the nearest hundredth of a meter.

The result of a calculation can't be more accurate than the least accurate number in the calculation.

The rules for working with significant digits are given below.

1. **Nonzero digits and zeros between nonzero digits are significant.**

245	3 significant digits
7.907	4 significant digits
907.08	5 significant digits

Lesson 6: Chemistry Computation

2. **Zeros to the left of the first nonzero digit in a number are *not* significant.**

   0.005   1 significant digit

   0.0739   3 significant digits

3. **Zeros at the end of a number to the right of the decimal point are significant.**

   12.000   5 significant digits

   0.580   3 significant digits

   1.0   2 significant digits

4. **Zeros at the end of a number greater than 1 and without decimal points are *not* significant.**

   1,200   2 significant digits

   10   1 significant digit

   100   1 significant digit

5. The coefficients of a balanced equation and numbers obtained by counting objects are infinitely significant. So if a balanced equation calls for 3 moles of carbon, we should think of it as 3.00 moles of carbon.

6. When multiplying and dividing, the answer should have the same number of significant digits as the number in the calculation with the least number of significant digits.

   $0.25 \times 60.335 = 15.08375 \rightarrow 15$

   (0.25 has 2 significant digits, so the answer can only have 2 significant digits.)

   $101 \times 12 = 1{,}212 \rightarrow 1200$

   (12 has 2 significant digits, so the answer can only have 2 significant digits.)

   $20 \div 5.023 = 3.98168\ldots \rightarrow 4$

   (20 has 1 significant digit, so the answer can only have 1 significant digit.)

7. When adding and subtracting, the answer should have the same number of decimal places as the number in the calculation with the least accurate number of decimal places.

   $26 + 45.88 + 0.09534 = 72$

   (26 has 2)

   $780 + 35 + 4 = 820$

Significant digits show the accuracy of a measurement or a calculation. Because the result of a calculation can't be more accurate than the least accurate number in the calculation, you must use the correct number of significant digits.

MOLECULAR WEIGHT

If you know the formula of a compound, you can find its molecular weight by adding up the atomic weights of all the atoms in the compound. *Remember*, you can find the atomic weight of any element by looking at the periodic table.

Find the molecular weight of carbon dioxide. First, you should know from the name that the formula of carbon dioxide is CO_2.

Molecular weight of CO_2 = (atomic weight of C) + 2(atomic weight of O)
Molecular weight of CO_2 = (12.0 amu) + 2(16.0 amu)
= 12.0 amu + 32.0 amu
= 44.0 amu

Try sodium nitrate, $NaNO_3$.

Molecular weight of $NaNO_3$ = (weight of Na) + (weight of N) + 3(weight of O)
Molecular weight of $NaNO_3$ = (23.0 amu) + (14.0 amu) + 3(16.0 amu)
= (23.0 amu) + (14.0 amu) + (48.0 amu)
= 85.0 amu

MOLES

If you have a dozen bagels, you have 12 bagels. *Dozen* is just a word for the number 12. In the same way, a **mole** is just a number. To have a mole of mercury atoms means you have a certain number of them. A mole happens to be a very large number, 6.02×10^{23}, and is also known as Avogadro's number. In chemistry, atoms are quantified in moles, and the mole serves as a bridge connecting all the different quantities that you'll come across in chemical calculations. The coefficients in chemical reactions give you the reactants and products in numbers of moles, and most of the stoichiometry (chemistry math) questions you'll see on the test will be exercises in converting between grams, liters, molecules, and moles.

MOLES AND ATOMS

The definition of Avogadro's number gives you the information you need to convert between moles and individual atoms and molecules:

1 mole = 6.02×10^{23} atoms

$$\text{Moles} = \text{atoms} \times \frac{1 \text{ mole}}{6.02 \times 10^{23} \text{ atoms}}$$

Try a conversion from atoms to moles. If you have 9.0×10^{23} of some atom (it doesn't matter what the atom is), how many moles do you have?

Use Dimensional Analysis to help you keep track of units. First, how do you relate moles and number of atoms to get a fraction that equals 1?

$$\frac{1 \text{ mole}}{6.02 \times 10^{23} \text{ atoms}} = 1$$

Now convert 9.0×10^{23} atoms to moles—

$$\text{Moles} = \text{atoms} \times \frac{1 \text{ mole}}{6.02 \times 10^{23} \text{ atoms}}$$

$$9.0 \times 10^{23} \text{ atoms} \times \frac{1 \text{ mole}}{6.02 \times 10^{23} \text{ atoms}} = 1.5 \text{ moles}$$

You can also convert from moles to atoms. If you have 3.0 moles of a substance, how many individual atoms are there?

$$\text{Moles} = \text{atoms} \times \frac{1 \text{ mole}}{6.02 \times 10^{23} \text{ atoms}}$$

Use algebra to get the expression below. Then plug in 3.0 for moles and solve the equation.

$$\text{Atoms} = (\text{moles}) \times \frac{6.02 \times 10^{23} \text{ atoms}}{1 \text{ mole}}$$

$$\text{Atoms} = 3.0 \text{ moles} \times \frac{6.02 \times 10^{23}}{1 \text{ mole}} = 18 \times 10^{23} \text{ atoms}$$

$$\text{Atoms} = 18 \times 10^{23} \text{ atoms} = 1.8 \times 10^{24} \text{ atoms}$$

Notice that the examples on the previous page followed the significant figure rules. In both cases, you should have an answer with two significant figures. Scientific notation makes reading significant figures easier for very large or very small numbers.

MOLES AND GRAMS

Moles and grams can be related using the atomic weights given in the periodic table. Atomic weights on the periodic table are given in terms of atomic mass units (amu), but an amu is the same as a gram per mole, so if 1 carbon atom weighs 12 amu, then 1 mole of carbon atoms weighs 12 grams.

You can use the relationship between amu and g/mol to convert between grams and moles by using the following equation:

$$\text{Moles} = \frac{\text{grams of substance}}{\text{atomic weight}}$$

Convert from grams to moles. If you have 10 grams of neon (Ne), how many moles do you have? In order to do this one, you look up the atomic weight of neon on the periodic table. It is 20.2 amu, which is the same as 20.2 grams per mole. Now use the formula.

$$\text{Moles} = \frac{\text{grams of substance}}{\text{atomic weight}}$$

$$\text{Moles} = \left(\frac{10 \text{ grams Ne}}{20.2 \text{ grams / mole}}\right) = 0.05 \text{ moles of Ne}$$

You can also convert from moles to grams. If you have 2.0 moles of phosphorous (P), how many grams do you have? Look up phosphorous on the periodic table to find its atomic weight, 31.0 grams/mole. Now solve the formula for grams.

$$\text{Moles} = \frac{\text{grams of substance}}{\text{atomic weight}}$$

Use algebra to rewrite the equation.

Grams = (moles) (atomic weight)

$$\text{Grams} = \left(\frac{2 \text{ moles of P}}{1}\right)\left(\frac{31.0 \text{ grams}}{1 \text{ mole}}\right) = 62 \text{ grams of P}$$

Lesson 6: Chemistry Computation

The two examples above show conversions between moles of atoms and grams. You can also convert between moles of molecules and grams. The only difference is that when you work with molecules, you have to find the molecular weight instead of the atomic weight. Say you have 180 grams of water (H_2O), and you want to know how many moles that is. Look at the periodic table to get the atomic weights of H (1 g/mole) and O (16 g/mole). Now figure out the molecular weight of water (2(1) + 16 = 18) and plug it in to the formula.

$$\text{Moles} = \frac{\text{grams of substance}}{\text{molecular weight}}$$

$$\text{Moles} = \left(\frac{180 \text{ grams of } H_2O}{1}\right)\left(\frac{1 \text{ mole}}{18.0 \text{ grams}}\right) = 10 \text{ moles of } H_2O$$

MOLES AND GASES

At STP, standard temperature and pressure, you can convert directly between moles of gas and liters of gas (at STP, $P = 1$ atmosphere and $T = 273$ K). That's because at STP, one mole of gas always occupies 22.4 liters of volume.

$$\boxed{\begin{array}{c} \text{For a gas at STP} \\ \text{Moles} = \dfrac{\text{liters}}{22.4 \text{ L/mole}} \end{array}}$$

Convert from liters of gas to moles. Say you have 11.2 liters of a gas (it can be any gas) at STP (the gas has to be at standard temperature and pressure for the conversion to work). Use the formula to get the number of moles of gas.

$$\text{Moles} = \frac{\text{liters}}{22.4 \text{ L/mole}}$$

$$\text{Moles} = \left(\frac{11.2 \text{ L}}{1}\right)\left(\frac{1 \text{ mole}}{22.4 \text{ L}}\right) = 0.5 \text{ moles}$$

You can also convert from moles to liters of gas. If you have 2.0 moles of gas at STP, how many liters of gas do you have? Just rearrange the formula so you can solve for liters.

$$\text{Moles} = \frac{\text{liters}}{22.4 \text{ L/mole}}$$

$$\text{Liters} = (\text{moles})(22.4 \text{ L/mole})$$

$$\text{Liters} = \left(\frac{2 \text{ moles}}{1}\right)\left(\frac{22.4 \text{ L}}{1 \text{ mole}}\right) = 44.8 \text{ liters of gas}$$

CHEMICAL EQUATIONS

A chemical equation relates the number of atoms or molecules involved in a reaction, not their masses. Here's a way to think about this: If you are making a bicycle, you need one frame and two wheels. Their weights don't matter, just their quantities. In the same way, you will want to keep track of the number of atoms that are part of a chemical reaction. For example, to make one molecule of water you need one atom of oxygen and two atoms of hydrogen. In chemistry, rather than counting atoms or molecules individually, you count them in moles simply because you can work with smaller numbers. You could also count bicycles in moles: one mole of frames and two moles of wheels would make one mole (6.02×10^{23}!) of bicycles.

The coefficients in a balanced chemical equation tell you the number of moles of each reactant and product involved in the reaction. The coefficients in a chemical equation do not directly tell you the mass or volume of the reactants and products. Look at a balanced equation.

$N_2 + 3 H_2 \rightarrow 2 NH_3$

The equation above tells you that 1 mole of N_2 will react with 3 moles of H_2 to form 2 moles of NH_3. What if you wanted to know how many moles of NH_3 are formed when 5 moles of N_2 react?

The best way to do this problem is to set up a ratio that compares the coefficients of the balanced equation (1 N_2 and 2 NH_3) to the molar quantities that you want to know about (5 moles of N_2 and x moles of NH_3).

$$\frac{N_2}{NH_3} = \frac{1}{2} = \frac{5}{x}$$

Lesson 6: Chemistry Computation

Now you can cross-multiply and solve for x.

x = 10 moles of NH_3

Here's another one.

$$2\ KClO_3 \rightarrow 2\ KCl + 3\ O_2$$

How many moles of KCl are produced in the reaction shown above if 12 moles of O_2 are produced?

First set up the ratio relating the coefficients in the balanced equation to the actual numbers of moles in the problem.

$$\frac{KCl}{O_2} = \frac{2}{3} = \frac{x}{12}$$

Now cross-multiply and solve for x.

x = 8 moles of KCl

Once you find the number of moles of a substance, you can find the grams or liters of the substance. You have to add another step to the problem. Look at the example below.

$$CH_4 + O_2 \rightarrow CO_2 + 2\ H_2O$$

If 2 moles of CH_4 are consumed in the reaction above, how many grams of water are produced?

First you have to find the number of moles of H_2O produced. You can do that by setting up the same kind of ratio that you used in the previous problems.

$$\frac{CH_4}{H_2O} = \frac{1}{2} = \frac{2}{x}$$

Now cross-multiply and solve for x.

x = 4 moles of H_2O

But you're not done yet. You have to convert 4 moles of H_2O into grams. You already know how to do that. From the periodic table, you know that the molecular weight of H_2O is 18 g/mole. Now use the formula to convert from moles to grams.

$$\text{Moles} = \frac{\text{grams}}{\text{molecular weight}}$$

Grams = (moles)(molecular weight)

Grams = (4 moles)(18 g/mole) = 72 grams of H_2O

So for this kind of problem, you have to do two steps. First, find the moles of water. Then, convert to grams of water. Try another one.

$$2\ NO(g) \rightarrow N_2(g) + O_2(g)$$

If 1.00 mole of NO gas is consumed in the reaction above, how many liters of N_2 gas are created at STP?

First, use the ratio to find the number of moles of N_2 gas created.

$$\frac{NO}{N_2} = \frac{2}{1} = \frac{1.00}{x}$$

Now, cross-multiply and solve for x.

$x = 0.500$ moles of N_2

Now that you know the number of moles of N_2, you can use the conversion formula to find the number of liters at STP.

$$\text{Moles} = \frac{\text{liters}}{(22.4\ L/mole)}$$

Liters = (moles)(22.4 L/mole)

Liters = (0.500 moles)(22.4 L/mole) = 11.2 liters of N_2

LIMITING REACTANTS

In a chemical reaction, you need all of the reactants to be present, if the reaction is to happen. Often, you will have different amounts of each reactant. The limiting reactant is the reactant that you run out of first. Once the limiting reactant is gone, the reaction stops. Look at the reaction below.

$$N_2 + 3\ H_2 \rightarrow 2\ NH_3$$

Say you have 1 mole of N_2 and 50 moles of H_2. Three moles of H_2 will react for every mole of N_2. Once the 1 mole of N_2 is gone, the reaction stops, even though you still have 47 moles of H_2 left. So N_2 is the limiting reactant in this case.

Say you had 3 moles of N_2 and 3 moles of H_2. When only 1 mole of N_2 has reacted, all 3 moles of H_2 will be gone, so you will run out of H_2 first in this case. Now, H_2 is the limiting reactant.

It's important to notice that the limiting reactant is not necessarily the reactant that you have the least amount of. If you had 1 mole of N_2 and 2 moles of H_2, you would still run out of H_2 first, and H_2 would be the limiting reactant, even though you started with more H_2 than N_2. That's because in this reaction, you need to have three times as much H_2 as N_2. If you have less than three times as much H_2, then you'll run out of H_2 first.

Percent Yield

In some of the chemical reaction problems above, you used balanced chemical equations to predict how many grams of products would be produced. When you use a balanced equation to predict the number of grams of product, you get a number called the **theoretical yield**.

In practice, when a chemical reaction occurs, you usually get fewer grams of products than you would have predicted from the balanced equation. The actual number of grams of a product is called the **actual yield**.

Some problems will ask for the **percent yield** of a reaction. That's a percentage that compares the actual yield to the theoretical yield. The formula for percent yield is shown below.

$$\text{Percent Yield} = \frac{\text{Actual Yield}}{\text{Theoretical Yield}} \times 100$$

So if you have a reaction with an actual yield of 22 grams of H_2O, and the theoretical yield predicted by the balanced equation is 25 grams, you can find the percent yield by using the formula.

$$\text{Percent Yield} = \frac{22 \text{ grams}}{25 \text{ grams}} \times 100 = 88\%$$

TIP: The closer the percent yield is to 100%, the closer the actual yield is to the theoretical yield.

REVIEW FOR LESSON 6

1. Which of the following shows the conversion of 0.43 m to centimeters?

   A $(0.43 \text{ m})\left(\dfrac{100 \text{ cm}}{1 \text{ m}}\right) =$

   B $(0.43 \text{ cm})\left(\dfrac{100 \text{ m}}{1 \text{ cm}}\right) =$

   C $(0.43 \text{ m})\left(\dfrac{100 \text{ cm}}{1 \text{ m}}\right)\left(\dfrac{1 \text{ m}}{100 \text{ cm}}\right) =$

   D $(0.43 \text{ m})\left(\dfrac{100 \text{ m}}{100 \text{ cm}}\right)\left(\dfrac{1 \text{ m}}{100 \text{ cm}}\right) =$

2. Which of the following shows the conversion of 650 g to kg?

   F $(650 \text{ g})\left(\dfrac{1{,}000 \text{ g}}{1 \text{ kg}}\right) =$

   G $(650 \text{ g})\left(\dfrac{10 \text{ g}}{1 \text{ kg}}\right) =$

   H $(650 \text{ g})\left(\dfrac{10 \text{ kg}}{1 \text{ g}}\right) =$

   J $(650 \text{ g})\left(\dfrac{1 \text{ kg}}{1{,}000 \text{ g}}\right) =$

3 Which of the following shows the conversion of 1.2 L to mL?

A $(1.2 \text{ L}) \left(\dfrac{1{,}000 \text{ mL}}{1 \text{ L}} \right) \left(\dfrac{1{,}000 \text{ mL}}{1 \text{ L}} \right) =$

B $(1.2 \text{ L}) \left(\dfrac{1{,}000 \text{ mL}}{1 \text{ L}} \right) \left(\dfrac{1 \text{ L}}{1{,}000 \text{ mL}} \right) =$

C $(1.2 \text{ L}) \left(\dfrac{1{,}000 \text{ mL}}{1 \text{ L}} \right) =$

D $(1.2 \text{ L}) \left(\dfrac{1 \text{ L}}{1{,}000 \text{ mL}} \right) =$

4 Which of the following measurements is the greatest?
F 10 mm
G 10 nm
H 10 μm
J 10 m

5 What is the value in liters of 340 milliliters?
A 0.34 L
B 3.4 L
C 34 L
D 3400 L

6 What is the value in grams of 0.15 kilograms?
F 1,500 g
G 150 g
H 1.5 g
J 0.015 g

7 Which of the following statements is true?
A One milliliter of water weighs one kilogram.
B One liter of water weighs one gram.
C One milliliter of water weighs one gram.
D One liter of water weighs one milligram.

8 A 95 gram sample of iron has a volume of 12 cm^3. What is the density of iron?

F 0.13 g/cm^3

G 7.9 g/cm^3

H 79 g/cm^3

J 1,100 g/cm^3

9 How many significant digits are in the measurement 1.76 meters?

A 1

B 2

C 3

D 4

10 How many significant digits are in the measurement 0.023 mg?

F 2

G 3

H 4

J 5

11 How many significant digits are in the measurement 2.00 L?

A 1

B 2

C 3

D 4

12 12 × 402 =

What is the answer to the multiplication problem above expressed with the proper number of significant digits?

F 5,000

G 4,800

H 4,820

J 4,824

Review for Lesson 6

13 1.72 + 3.1 =

What is the answer to the addition problem above expressed with the proper number of significant digits?

A 4.82
B 4.8
C 5.0
D 5

14 What is the molecular weight of silicon dioxide?

F 16.0 g/mole
G 32.0 g/mole
H 44.1 g/mole
J 60.1 g/mole

15 What is the weight of one mole of copper(I) chloride?

A 99.0 grams
B 134.5 grams
C 162.5 grams
D 198.0 grams

16 What is the weight of 6.02×10^{23} molecules of boron trichloride?

F 81.8 grams
G 117.3 grams
H 128.1 grams
J 152.8 grams

17. A tank contains 33.6 liters of helium gas at standard temperature and pressure. How many moles of He gas are in the tank?

   A 0.5 moles
   B 1.0 moles
   C 1.5 moles
   D 2.0 moles

18. A tank contains 33.6 liters of helium gas at standard temperature and pressure. How many grams of He gas are in the tank?

   F 2.0 grams
   G 4.0 grams
   H 6.0 grams
   J 8.0 grams

19. How many molecules are contained in a 1.35-mole sample of CO_2?

   A 1.35×10^{23} molecules
   B 6.02×10^{23} molecules
   C 7.37×10^{23} molecules
   D 8.13×10^{23} molecules

20. How many moles of $CaCO_3$ are contained in a pure sample that weighs 961 grams?

   F 2.40 moles
   G 4.80 moles
   H 7.20 moles
   J 9.60 moles

21. What is the weight in grams of 5.00 moles of solid iron(Fe)?

   A 55.8 grams
   B 112 grams
   C 225 grams
   D 279 grams

22. What is the volume occupied by 3.5 moles of nitrogen gas at STP?

   F 14.4 L
   G 49.0 L
   H 78.4 L
   J 98.0 L

Review for Lesson 6

23. What is the mass of a sample of ammonia (NH_3) gas that occupies 44.8 liters at STP?
   A 34.0 grams
   B 22.4 grams
   C 17.0 grams
   D 11.2 grams

24.
$$2\ Al + 3\ Cl_2 \rightarrow 2\ AlCl_3$$

   If 3 moles of $AlCl_3$ were produced in the reaction above, how many moles of Cl_2 were consumed?
   F 1.5
   G 3.0
   H 4.5
   J 6.0

25. The reaction below took place at STP.

$$C_3H_8(g) + 5\ O_2(g) \rightarrow 3\ CO_2(g) + 4\ H_2O(g)$$

   How many liters of CO_2 gas were produced if 0.5 moles of C_3H_8 were consumed in the reaction?
   A 11.2 L
   B 33.6 L
   C 44.8 L
   D 57.0 L

26.
$$Be + 2\ HCl \rightarrow BeCl_2 + H_2$$

   If 6.0 moles of HCl were consumed in the reaction above, what was the mass of beryllium consumed in the reaction?
   F 3.0 grams
   G 9.0 grams
   H 18 grams
   J 27 grams

27.
$$ZnSO_3 \rightarrow ZnO + SO_2$$

   How many grams of ZnO are produced when 5.00 moles of $ZnSO_3$ are consumed in the reaction above?
   A 81.4 grams
   B 163 grams
   C 407 grams
   D 814 grams

28

$$4\ NH_3 + 5\ O_2 \rightarrow 4\ NO + 6\ H_2O$$

If 12 moles of NO are produced in the reaction above, how many moles of H_2O are produced?

F 18
G 15
H 12
J 9

29

$$CH_4 + 2\ O_2 \rightarrow CO_2 + 2\ H_2O$$

In an experiment, 2 moles of CH_4 were placed in a container with 2 moles of O_2. The reaction above took place, producing 1 mole of CO_2 and 2 moles of H_2O. Which of the following was the limiting reactant?

A CH_4
B O_2
C CO_2
D H_2O

30 In a commercial process that produces ammonia, the theoretical yield of NH_3 was found to be 150 grams. When the process was carried out, the actual yield was 135 grams. What was the percent yield?

F 10%
G 50%
H 80%
J 90%

Review for Lesson 6

ANSWERS AND EXPLANATIONS

1. **A is correct.** In choice **A**, 0.43 m is multiplied by a fraction equal to 1, with meters in the denominator so that the only units left at the end will be cm. By the way, 0.43 m is equal to 43 cm.

2. **J is correct.** In choice **J**, 650 g is multiplied by a fraction equal to 1, with grams in the denominator so that the only units left at the end will be kg. By the way, 650 g is equal to 0.65 kg.

3. **C is correct.** In choice **C**, 1.2 L is multiplied by a fraction equal to 1, with L in the denominator so that the only units left at the end will be mL. So 1.2 L is equal to 1,200 mL.

4. **J is correct.** Answer choices **F**, **G**, and **H** are smaller units than meters. Choice **J**, 10 m, is the greatest.

5. **A is correct.** The expression below shows how to do the conversion.

$$(340 \text{ mL})\left(\frac{1 \text{ L}}{1,000 \text{ mL}}\right) = \left(\frac{340}{1,000}\right) \text{L} = 0.34 \text{ L}$$

6. **G is correct.** The expression below shows how to do the conversion.

$$(0.15 \text{ kg})\left(\frac{1,000 \text{ g}}{1 \text{ kg}}\right) = (0.15)(1,000) \text{ g} = 150 \text{ g}$$

7. **C is correct.** One milliliter of water weighs one gram. By the way, one milliliter of water is also the same as one cm^3 (cubic centimeter).

8. **G is correct.** You can use the following formula to calculate density.

$$\text{Density} = \frac{m}{V} = \frac{95 \text{ grams}}{12 \text{ cm}^3} = 7.9 \text{ g}/\text{cm}^3$$

9. **C is correct.** 1.76 has 3 digits, and all of them are significant.

10. **F is correct.** The zeros to the left of the first nonzero digit are not significant, so 0.023 mg has two significant digits.

11. **C is correct.** Zeros at the end of a number to the right of the decimal point are significant, so 2.00 L has three significant digits.

12 **G is correct.** 12 has two significant digits, so those are the most significant digits that can appear in the answer. 4,824 must be rounded to 4,800.

13 **B is correct.** 3.1 has only one decimal place, so that's the most decimal places that can appear in the answer. 4.82 must be rounded to 4.8.

14 **J is correct.** The formula for silicon dioxide is SiO_2. From the periodic table, the atomic weight of Si is 28.1 and the atomic weight of O is 16.0. The units for atomic weight can be either amu or grams/mole.

$$28.1 \text{ g/mole} + 2(16.0 \text{ g/mole}) = 60.1 \text{ grams/mole}$$

15 **A is correct.** The molecular weight of a compound is the weight of one mole, so this question just asks for the molecular weight. The formula for copper(I) chloride is CuCl. From the periodic table, the atomic weight of Cu is 63.5 g/mole and the atomic weight of Cl is 35.5 g/mole.

$$63.5 \text{ grams} + 35.5 \text{ grams} = 99.0 \text{ grams}$$

16 **G is correct.** 6.02×10^{23} molecules is the same as 1 mole, so all you have to do is find the molecular weight of the compound. The formula for boron trichloride is BCl_3. From the periodic table, the atomic weight of B is 10.8 g/mole and the atomic weight of Cl is 35.5 g/mole.

$$10.8 \text{ grams} + 3(35.5 \text{ grams}) = 117.3 \text{ grams}$$

17 **C is correct.** Use the formula that converts from liters to moles at STP.

$$\text{Moles} = \frac{\text{liters}}{(22.4 \text{ L/mole})}$$

$$\text{Moles of He} = \frac{33.6 \text{ L}}{(22.4 \text{ L/mole})} = 1.5 \text{ moles of He}$$

18 **H is correct.** This is a two-step problem. To convert from liters to grams, first you have to convert from liters to moles, and then from moles to grams.

Review for Lesson 6

$$\text{Moles} = \frac{\text{liters}}{(22.4 \text{ L/mole})} = \frac{33.6 \text{ L}}{(22.4 \text{ L/mole})} = 1.5 \text{ moles of He}$$

You can find the atomic weight of He (4.0 g/mole) on the periodic table.

$$\text{Grams} = \text{Moles} = \times \text{(atomic weight)}$$

$$\text{Grams} = 1.5 \text{ moles} \left(\frac{4.0 \text{ g}}{\text{mole}}\right) = 6.0 \text{ grams of He}$$

19 D is correct. Use the formula to convert from moles to molecules. Notice that it doesn't matter what molecule you are working with, so you never use any information specific to CO_2.

$$\text{Moles} = \frac{\text{molecules}}{(6.02 \times 10^{23})}$$

Molecules = (moles)(6.02 × 10^{23} molecules/mole)

Molecules = (1.35 moles)(6.02 × 10^{23} molecules/mole)

Molecules = 8.13 × 10^{23} molecules

You can use POE to eliminate choices **F** and **G** if you notice that the answer must be greater than 1 mole, so it must be greater than 6.02 × 10^{23}.

20 J is correct. First, you need to find the molecular weight of $CaCO_3$. Then, you can use the formula to convert from grams to moles. From the periodic table, the atomic weights are: Ca = 40.1, C = 12.0, and O = 16.0.

$$40.1 \text{ g/mole} + 12.0 \text{ g/mole} + 3(16.0 \text{ g/mole}) = 100.1 \text{ g/mole}$$

Now use the formula.

$$\text{Moles} = \frac{\text{grams}}{\text{molecular weight}}$$

$$\text{Moles} = \frac{961 \text{ grams}}{100.1 \text{ g/mole}} = 9.60 \text{ moles}$$

21 **D is correct.** From the periodic table, the atomic weight of Fe is 55.8 g/mole. Now we can use the formula to convert moles to grams.

$$\text{Moles} = \frac{\text{grams}}{\text{atomic weight}}$$

Grams = (moles)(atomic weight)

Grams = (5.00 moles)(55.8 g/mole) = 279 grams

22 **H is correct.** You can use the formula to convert from moles of gas to liters. Notice that the fact that the gas is nitrogen doesn't matter in the calculation.

$$\text{Moles} = \frac{\text{liters}}{22.4 \text{ L/mole}}$$

Liters = (moles)(22.4 L/mole)

Liters = (3.5 moles)(22.4 L/mole) = 78.4 liters

23 **A is correct.** This is a two-part problem. First, you have to convert from liters of gas to moles. Then, you can convert from moles to grams. Use the following formulas.

$$\text{Moles} = \frac{\text{liters}}{22.4 \text{ L/mole}}$$

Grams = moles × (molecular weight)

To calculate grams, you need to get the molecular weight of NH_3. From the periodic table, the atomic weight of N is 14.0 g/mole and the atomic weight of H is 1.0 g/mole.

Molecular weight (NH_3) = 14.0 g/mole + 3 (1.0 g/mole) = 17.0 g/mole

Now use the formulas.

$$\text{Moles} = \left(\frac{44.8 \text{ L}}{22.4 \text{ L/mole}}\right) = 2 \text{ moles}$$

$$\text{Grams} = 2 \text{ moles} \times \left(\frac{17.0 \text{ g}}{\text{mole}}\right) = 34.0 \text{ grams NH}_3$$

24 H is correct. Set up a ratio that compares the coefficients of the balanced equation (3 Cl_2 and 2 $AlCl_3$) to the molar quantities that you want to know about (x moles of Cl_2 and 3 moles of $AlCl_3$).

$$\frac{Cl_2}{AlCl_3} = \frac{3}{2} = \frac{x}{3}$$

$x = 4.5$ moles of Cl_2

25 B is correct. This is a two-step problem. First, you should set up a ratio that compares the coefficients of the balanced equation (1 C_3H_8 and 3 CO_2) to the molar quantities that you want to know about (0.5 moles of C_3H_8 and x moles of CO_2).

$$\frac{C_3H_8}{CO_2} = \frac{1}{3} = \frac{0.5}{x}$$

$x = 1.5$ moles of CO_2

Now you can use the conversion formula to convert moles of CO_2 to liters of CO_2.

$$Moles = \frac{liters}{22.4 \text{ L/mole}}$$

Liters = (moles)(22.4 L/mole)

Liters = (1.5 moles)(22.4 L/mole) = 33.6 liters of CO_2

26 **J is correct.** This is a two-step problem. First, you should set up a ratio that compares the coefficients of the balanced equation (1 Be and 2 HCl) to the molar quantities that you want to know about (x moles of Be and 6 moles of HCl).

$$\frac{Be}{HCl} = \frac{1}{2} = \frac{x}{6.0}$$

x = 3.0 moles of Be

Now you can use the conversion formula to convert moles of Be to grams of Be. From the periodic table, you know that the atomic weight of Be is 9.0 grams/mole.

$$Moles = \frac{grams}{atomic \text{ } weight}$$

Grams = (moles)(atomic weight)

Grams = (3.0 moles)(9.0 g/mole) = 27 grams of Be

27 **C is correct.** For this problem, you don't need to set up a ratio of coefficients because all of the coefficients are equal to 1. So if 5.00 moles of $ZnSO_3$ are consumed, then 5.00 moles of ZnO are produced. You can look at the periodic table to find the molecular weight of ZnO and then convert from moles to grams.

$$65.4 \text{ g/mole} + 16.0 \text{ g/mole} = 81.4 \text{ g/mole}$$

$$Moles = \frac{grams}{atomic \text{ } weight}$$

Review for Lesson 6

Grams = (moles)(atomic weight)

Grams = (5.00 moles)(81.4 g/mole) = 407 grams of ZnO

28 F is correct. Set up a ratio that compares the coefficients of the balanced equation (4 NO and 6 H_2O) to the molar quantities that you want to know about (12 moles of NO and x moles of H_2O).

$$\frac{NO}{H_2O} = \frac{4}{6} = \frac{12}{x}$$

$x = 18$ moles of H_2O

29 B is correct. The reaction requires twice as much O_2 as CH_4, so if the two compounds are present in equal quantities, you will run out of O_2 first. That makes O_2 the limiting reactant. Choices **C** and **D** can't be right because CO_2 and H_2O are products, not reactants.

30 J is correct. You can use the formula for percent yield.

$$\text{Percent Yield} = \frac{\text{Actual Yield}}{\text{Theoretical Yield}} \times 100$$

$$\text{Percent Yield} = \frac{135 \text{ grams}}{150 \text{ grams}} \times 100 = 90\%$$

You can use POE to eliminate choices **A** and **B** because 135 is greater than 50% of 150.

LESSON 7
GASES

MORE UNITS

TEMPERATURE

Scientists use two scales, the Celsius scale and the Kelvin scale, to measure temperature. The Celsius scale was devised so that the freezing point of water is 0°C and the boiling point of water is 100°C. Between those two reference points are 100 units called degrees Celsius.

On the Kelvin scale the freezing point of water is about 273 K and the boiling point of water is about 373 K. Notice how the Kelvin scale uses the same temperature difference as Celsius. One degree Celsius is equal to one Kelvin. This temperature scale was devised so that the 0 K represents absolute zero, or the lowest temperature possible. This occurs at −273°C.

Convert between the two temperature scales by adding or subtracting 273.

$$\text{Kelvin} = °\text{Celsius} + 273$$

$$°\text{C} = \text{K} - 273$$

Here are some things you should know:

- Absolute zero is 0 K or −273°C. At this temperature, all motion stops.
- The freezing point of water is 273 K or 0°C.
- The boiling point of water is 373 K or 100°C.

Note: You must *always* use Kelvin when you do calculations involving gases.

PRESSURE

In chemistry, pressure is measured in atmospheres (atm) or millimeters of mercury (mm Hg or torr). You can convert between these units using the following expression.

   1 atm = 760 mm Hg

You can use either of these units when you do calculations involving gases, as long as you consistently use the same unit throughout the problem.

Lesson 7: Gases

STP

STP stands for Standard Temperature and Pressure. Because the temperature, volume, and pressure of a gas are dependent on each other, scientists agreed upon STP as a useful reference. Make yourself familiar with STP, which comes up fairly often in problems involving gases. Assume the STP values below apply in a problem unless you're given different values.

At STP: Pressure = 1 atmosphere or 760 millimeters of mercury (mm Hg)

             Temperature = 0°C or 273 K

KINETIC MOLECULAR THEORY

The kinetic molecular theory, also known as the kinetic theory of gases, describes the behaviors of gases based on their molecular properties. This theory assumes that all gases display similar behaviors regardless of their chemical natures. In other words, the theory assumes that substances act as ideal gases.

Ideal gases have the following characteristics:

1. Temperature is directly proportional to the average **kinetic energy** of a gas. Kinetic energy is represented by the equation below.

$KE = \frac{1}{2} mv^2$

 m is the mass and v is the velocity of the molecule. Gas moves randomly in all directions. Because velocity is made up of speed and direction of motion, different particles of the same gas will have different kinetic energies. Temperature is a measure of the average kinetic energy of the gas particles. So, at the same temperature all gases will have the same average kinetic energy.

2. The volume of gas molecules is very small compared to the volume of the container. So, the volume of the gas molecules is insignificant when studying gases.

3. There are no forces of attraction or repulsion between the gas molecules. Because the gas molecules are very small compared to the volume of the container, they are relatively far apart from each other and do not affect other molecules.

4. Collisions between gas molecules are elastic. During an **elastic collision**, no kinetic energy is lost. Think of the molecules colliding and bouncing off each other without losing speed or mass.

THE GAS LAWS

The behavior of gases described by the kinetic theory is affected by four factors—the pressure, volume, and temperature of the gas sample, and the number of moles of gas. These factors are interrelated and influence each other.

BOYLE'S LAW

For example, think of a balloon filled with gas at room temperature. Squeezing the balloon decreases the volume of the gas and increases the pressure of the gas. Remember, the molecules of gas are moving randomly inside the volume of the balloon. As the volume decreases, the same particles move around a smaller space, exerting more pressure on the walls of the balloon. The number of particles and the temperature does not change. The opposite is also true, if you stop squeezing the balloon and increase the volume, the pressure will decrease. Because the container volume is now larger, the gas molecules will strike the wall of the balloon less often. This results in a decreased pressure. The relationship between pressure and volume is described by Boyle's Law. For a given amount of gas at a constant temperature, **Boyle's law** states that volume varies inversely with pressure.

$$P_1 V_1 = P_2 V_2$$

CHARLES'S LAW

At a constant pressure, gas volume is influenced by the temperature of the gas. Remember what temperature indicates about the gas molecules. An increase in temperature corresponds to an increase in average kinetic energy of the gas molecules. As the gas molecules move around faster, they strike the walls of the container with greater frequency and force. For pressure to remain constant, the walls of the container must move farther apart so that the gas molecules strike the walls with the same frequency and force. This relationship is summed up by Charles's law. For a given amount of gas at a constant pressure, **Charles's law** states that the volume varies directly with temperature.

$$\frac{V_1}{T_1} = \frac{V_2}{T_2}$$

> **TIP:** All gas laws are based on a Kelvin scale for temperature. **Always** use Kelvin for temperature. If temperature is given in Celsius, add 273 to get the temperature in Kelvin.

GAY-LUSSAC'S LAW

Finally, there is a relationship between the pressure and temperature of a gas when volume is held constant. When the temperature of a gas increases, the gas molecules move faster and strike the walls of the container with greater frequency and force. If volume is held constant, the pressure of the gas increases. This relationship is summed up by Gay-Lussac's Law. For a given amount of gas at a constant volume, **Gay-Lussac's law** states that the pressure varies directly with temperature.

$$\frac{P_1}{T_1} = \frac{P_2}{T_2}$$

THE COMBINED GAS LAW

The **combined gas law** unites all of the previous laws you just learned into one equation. This law relates pressure, volume, and temperature for a given number of gas molecules. Notice that when one of the factors is held constant, it will cancel out of the following equation and result in one of the earlier gas laws.

$$\frac{P_1 V_1}{T_1} = \frac{P_2 V_2}{T_2}$$

THE IDEAL GAS LAW

Nearly all behaviors of gas can be explained by one relationship, the ideal gas law. The **ideal gas law** relates all of the factors influencing gas behavior: pressure, volume, temperature, and the number of moles of a gas. R is the ideal gas constant, an experimentally determined value. If you know three out of the four variables, you can find the fourth.

$$PV = nRT$$

n is the number of moles of gas and R is the gas constant. The gas constant is a numerical value equal to $0.0821 \frac{L \cdot atm}{mol \cdot K}$. There are other values of R depending on the units.

OTHER GAS LAWS

AVOGADRO'S LAW
Remember you learned that a mole of gas will occupy 22.4 liters at STP. This was discovered by Avogadro, who determined the relationship between the number of particles of a gas and its volume. This relationship is constant regardless of the identity and size of the gas molecules. For a given temperature and pressure, **Avogadro's law** states that the number of gas particles is directly proportional to its volume. Equal moles of gas occupy equal volumes, and equal volumes of gas contain equal moles.

GRAHAM'S LAW
Graham's law describes the movement of gas particles during diffusion and effusion. **Diffusion** is the movement of gas particles from areas of high concentration to areas of low concentrations. **Effusion** is the movement of gas particles out of a small hole. Recall that there are no forces of attraction or repulsion between different gas molecules, so particles in space will flow randomly around. At a constant temperature, the particles have the same average kinetic energy (KE = $\frac{1}{2}mv^2$). When there is a mix of particles with different masses, the heavier particles move slower than the lighter particles.

Graham's law shows that diffusion of a heavier gas is slower than diffusion of a lighter gas. The law also shows that effusion of a heavier gas is slower than effusion of a lighter gas.

Lesson 7: Gases

DALTON'S LAW

Dalton's Law states that the total pressure of a mixture of gases is the sum of all the pressures of the individual gases in the mixture. The pressure exerted by each single gas in the mixture is called its partial pressure. For example, if there is a mixture of gas A and gas B, the total pressure of the mixture is the sum of the partial pressure of gas A and the partial pressure of gas B. The partial pressure of a gas is independent of its identity.

$$P_{total} = P_a + P_b + \ldots P_n$$

Dalton's law of partial pressures can also be used to determine the molar quantities of the gases in a mixture. The partial pressure of a gas over the total pressure of the mixture is equal to the molar fraction of that gas in the mixture. This can be shown in the equation below.

$$\frac{P_a}{P_{total}} = \frac{Moles_a}{Total\ moles}$$

REVIEW FOR LESSON 7

1. The boiling point of water is
   A 0°C
   B 273°C
   C 273 K
   D 373 K

2. The temperature and pressure at STP are
   F 0 K and 0 mm Hg
   G 0°C and 0 mm Hg
   H 273 K and 760 mm Hg
   J 273°C and 760 mm Hg

3. Normal room temperature is 25°C. What is this temperature expressed in Kelvin?
   A 25 K
   B 248 K
   C 298 K
   D 398 K

4. A student measured the pressure of the gas contained in a tank at 646 mm Hg. What is this pressure in atmospheres?
   F 0.85 atm
   G 1 atm
   H 1.2 atm
   J 646 atm

5. Which of the following statements is *not* true of an ideal gas?
   A The volume of the gas molecules is insignificant.
   B The pressure of the gas is always equal to one atmosphere.
   C There are no attractive forces between the individual gas molecules.
   D The kinetic energy of the gas molecules is proportional to the temperature in Kelvin.

6. A 2.0 mole sample of gas is at a temperature of 20°C and a pressure of 1.2 atm. What is the volume occupied by the gas?
   (R = the gas constant, 0.0821 L-atm/mol-K)
   F 20 L
   G 30 L
   H 40 L
   J 50 L

7. A helium balloon with a volume of 3.0 liters is at a temperature of 27°C and a pressure of 1.0 atm. How many moles of helium are in the balloon?
   A 0.12 moles
   B 0.60 moles
   C 1.2 moles
   D 6.0 moles

8 If the volume of a sample of ideal gas is doubled at constant temperature, the pressure of the gas will—

F double
G remain the same
H be half the initial pressure
J be reduced by a factor of four

9 Oxygen gas is kept in a 6-liter tank at a pressure of 740 mm Hg and a temperature of 20°C. If the temperature is increased to 60°C, what will be the new pressure in the tank?

A 746 mm Hg
B 780 mm Hg
C 841 mm Hg
D 2,220 mm Hg

10 A balloon with an initial volume of 32.0 L is rising through the atmosphere. As the balloon rises, the pressure decreases from 760 mm Hg to 710 mm Hg. The temperature decreases from 298 K to 294 K. What is the volume of the balloon under the new conditions?

F 34.7 L
G 34.2 L
H 33.8 L
J 29.5 L

11 As a gas is heated at a constant pressure, it will expand and take up more volume. If a 6-liter sample of gas is initially at a temperature of 10°C, what will be its volume if it is heated to 30°C?

A 6.4 L
B 9.0 L
C 16 L
D 18 L

12 A chamber contains equal molar quantities of N_2, O_2, and He gas. If the total pressure in the container is 768 mm Hg, what is the partial pressure due to O_2 gas?

F 768 mm Hg
G 384 mm Hg
H 256 mm Hg
J 192 mm Hg

13 Gas sample A occupies twice as much volume as gas sample B at the same pressure and temperature. Which of the following statements is true about the two gas samples?

A Sample A has half the molecular weight of sample B.
B Sample A has twice the molecular weight of sample B.
C Sample A has half as many moles as sample B.
D Sample A has twice as many moles as sample B.

14 Which of the following statements is always true about two different gas samples at the same temperature?

F They have the same average kinetic energy.

G They have the same molecular weight.

H They occupy the same volume.

J They exert the same pressure.

15 A tank contains equal molar quantities of He, Ne, Ar, and Kr. If a tiny valve is opened, which gas will escape from the tank the fastest?

A He

B Ne

C Ar

D Kr

Review for Lesson 7

ANSWERS AND EXPLANATIONS

1. **D is correct.** The boiling point of water is either 373 K or 100°C.

2. **H is correct.** The temperature at STP is either 273 K or 0°C. The pressure at STP is either 760 mm Hg or 1 atm. If you know *either* the temperature *or* the pressure, but not both, you can still use POE to eliminate two wrong answer choices.

3. **C is correct.** You can use the conversion formula.

$$K = \_\_°C + 273$$

$$K = 25 + 273 = 298$$

4. **F is correct.** You can convert because you know that 1 atm = 760 mm Hg.

$$646 \text{ mm Hg} \; \frac{1 \text{ atm}}{760 \text{ mm Hg}} = \left(\frac{646}{760}\right) \text{atm} = 0.85 \text{ atm}$$

You can use POE to answer this one if you realize that the answer must be less than 1 atm, and only one of the answer choices is less than 1 atm.

5. **B is correct.** An ideal gas can have any pressure, not just one atmosphere. All of the other choices are basic assumptions that we make about ideal gases. Use Process of Elimination to get rid of answers that describe ideal gases.

6. **H is correct.** Use $PV = nRT$. First solve for V, then plug in the numbers. Don't forget to convert from degrees Celsius to Kelvin (K = 20 + 273 = 293).

$$V = \frac{nRT}{P} = \frac{(2.0)(0.0821)(293)}{(1.2)} \text{ L} = 40 \text{ L}$$

7. **A is correct.** Use $PV = nRT$. First solve for n, then plug in the numbers. Don't forget to convert from degrees Celsius to Kelvin (K = 27 + 273 = 300).

$$n = \frac{PV}{RT} = \frac{(1.0)(3.0)}{(0.0821)(300)} \text{ moles} = 0.12 \text{ moles}$$

8 H is correct. Use Boyle's law.

$$P_1 V_1 = P_2 V_2$$

If the V_2 is twice as big as V_1, then P_2 must be half as big as P_1 for the equation to work.

9 C is correct. Because the volume doesn't change, you can ignore it and use the formula that compares pressure and temperature. Don't forget to convert degrees Celsius to Kelvin (20 + 273 = 293; 60 + 273 = 333)

$$\frac{P_1}{T_1} = \frac{P_2}{T_2}$$

$$\frac{740 \text{ mm Hg}}{293 \text{ K}} = \frac{P_2}{333 \text{ K}}$$

$$P_2 = \left(\frac{740}{293}\right)(333) \text{ mm Hg} = 841 \text{ mm Hg}$$

10 H is correct. Use the formula that compares pressure, volume, and temperature.

$$\frac{P_1 V_1}{T_1} = \frac{P_2 V_2}{T_2}$$

$$\frac{(760 \text{ mm Hg})(32.0 \text{ L})}{(298 \text{ K})} = \frac{(710 \text{ mm Hg})(V_2)}{(294 \text{ K})}$$

$$V_2 = (32.0 \text{ L})\left(\frac{760}{710}\right)\left(\frac{294}{298}\right) = 33.8 \text{ L}$$

11 A is correct. Because the pressure doesn't change, you can ignore it and use Charles's law. Don't forget to convert degrees Celsius to Kelvin (10 + 273 = 283; 30 + 273 = 303).

$$\frac{V_1}{T_1} = \frac{V_2}{T_2}$$

$$\frac{6.0 \text{ L}}{283 \text{ K}} = \frac{V_2}{303 \text{ K}}$$

$$P_2 = \left(\frac{6.0}{283}\right)(303) \text{ mm Hg} = 6.4 \text{ L}$$

12 H is correct. According to Dalton's law, the partial pressure of a gas is directly proportional to the number of moles of the gas in the container. One-third of the moles of gas are He in the container, so one-third of the partial pressure must be from He.

$$(768 \text{ mm Hg})\left(\frac{1}{3}\right) = 256 \text{ mm Hg}$$

13 D is correct. According to Avogadro's law, the volume of a gas is directly proportional to the number of moles. A gas that occupies twice as much volume as another gas under the same conditions must have twice as many moles. The volume of a gas doesn't tell anything about the molecular weight.

14 F is correct. The temperature of a gas is a measure of its average kinetic energy, so two gas samples at the same temperature will have the same kinetic energy. If you only know the temperature of a gas, you can't tell anything about the number of moles, volume, or pressure.

15 A is correct. According to Graham's law, if different gases are at the same temperature, the lightest gas will move the fastest. Helium, the lightest of the gases listed, will move the fastest and escape from the tank the fastest.

LESSON 8
PHASE CHANGES

PHASES OF MATTER

Matter can undergo physical changes as well as chemical changes. Melting, freezing, and boiling are all examples of physical changes. Remember that physical changes don't make or break bonds, instead they affect the interactions between molecules or atoms. For example, melting ice into water does not change the molecules of H_2O into something else, it only changes the way the molecules are packed together.

Another idea to keep in mind is that physical changes are closely related to temperature. What does temperature tell us about matter? Temperature is directly related to the amount of energy molecules have and the level of organization among the molecules. The third idea to consider here is that every type of matter experiences intermolecular forces. That is, the individual molecules have some degree of stickiness toward each other. Every type of matter is different, so the strength of intermolecular forces varies quite a bit. For example, gold, hydrogen gas, and water all have different strengths of intermolecular forces. These differences give rise to the different melting and boiling points of gold, hydrogen gas, and water.

If we increase the temperature at a given pressure, matter typically goes from solid to liquid to gas. As the temperature increases, the matter absorbs energy in the form of kinetic energy. Remember that kinetic energy is energy of motion, and, as motion increases, the level of organization of molecules decreases. For example, solid water, or ice, is highly organized. The molecules of H_2O are packed in a very orderly way, and they don't move very much. If you increase the temperature of the ice, the molecules will eventually absorb enough energy to move around and the organized structure of the molecules will break down. That is, the ice will melt into liquid water. Liquid H_2O has more energy and less organization than solid ice. Likewise, if you continue to increase the temperature, the liquid H_2O absorbs even more energy. You know this will cause an increase in the movement, or kinetic energy, of the molecules. Eventually, the molecules will have so much energy and motion that they will escape interactions with each other and float away as a vapor. That is, the water will boil into a gas. How do the kinetic energy and organization of water vapor compare with liquid water? Gaseous H_2O has more kinetic energy, so it has more motion and less organization than liquid H_2O.

SOLIDS

Solids have the least kinetic energy and greatest organization of all the phases. Matter exists in a solid phase at a relatively low temperature. Molecules or ions in a solid are not free to move very much. What holds the molecules together? The interactions between the molecules. Although the strength of the interactions will vary greatly between types of matter, all types of substances have some level of intermolecular forces that cause the molecules to stick to each other.

LIQUIDS

If a solid is heated, the molecules will eventually absorb enough energy to overcome some of the intermolecular forces and begin to move. You recognize this as melting. The atoms that make up the molecules are still bonded together, but the individual molecules are more loosely packed together. Liquids have more kinetic energy than solids have, and they are less organized. Because of the differences among the intermolecular forces in different types of matter, the melting point of each substance is unique.

The **viscosity** of a liquid is its thickness and stickiness. The more tightly the molecules or ions of a liquid stick to each other and to the walls of its container, the more viscous the liquid is. For instance, maple syrup is more viscous than water.

All liquids have **surface tension**. Surface tension is what makes some liquids "bead up" on a flat surface. The stronger the bonds in the liquid, the more surface tension it will have and the more tightly it will form beads.

GASES

If a liquid is heated, *all* of the molecules will eventually have enough motion to overcome the intermolecular forces and become a vapor. In a gas, all of the molecules are free to move independently, so a gas is the most energetic and the least organized phase of matter. For the same reason that melting points vary among different types of matter, boiling points vary as well: Different amounts of energy are needed to disrupt various intermolecular forces.

At extremely high temperatures, matter exists in a gas-like phase where atoms are stripped of their electrons. This collection of ions and electrons is called **plasma**.

ATTRACTIVE FORCES IN SOLIDS AND LIQUIDS

As discussed earlier, all matter has some intermolecular attraction, and these forces are what hold a substance in a solid or a liquid phase. There are many types of intermolecular forces. Here's the rundown:

IONIC BONDS

An ionic solid is held together by attractions of positive and negative ions, also called the **electrostatic force**. The NaCl structure is an extended, three-dimensional lattice in which every sodium cation interacts with several chloride anions, and vice versa. That is, individual NaCl molecules don't exist. Because the electrostatic force is a very strong type of interaction, ionic bonds are strong and substances held together by ionic bonds have high melting and boiling points.

Conduction of electricity requires the free movement of electrons. Because the electrons in ionic solids are held tightly by the ions, electrons do not move around the solid, and, therefore, ionic solids are poor conductors of electricity. But what about an ionic compound that has been melted into a liquid? In this case, the ions themselves are free to move around and so ionic liquids do conduct electricity.

What types of compounds are held together by ionic bonds? Here's quick review: Elements whose electronegativities are very different tend to form ionic bonds. Rather than sharing an electron pair to gain a full octet, ionic solids will completely transfer electrons from one atom to another to obtain a valence octet. Salts are held together by ionic bonds. Some examples are KF, NaBr, $CaCl_2$, and MgO. You can probably think of several more by simply looking at the periodic table for the elements with very different electronegativities.

NETWORK (COVALENT) BONDS

In a network solid, atoms are held together in a network of covalent bonds and individual molecules do not exist. The formula used for network solids is the empirical formula, which gives the ratio of elements in a compound. For example, quartz is made of SiO_2, but solid quartz consists of an extended network of covalently bonded Si and O atoms in 1:2 ratio. You can visualize a network solid as one big molecule. Network solids are very hard and have very high melting and boiling points.

The electrons in a network solid are located in covalent bonds between particular atoms, so they are not free to move about the lattice. This makes network solids poor conductors of electricity.

The most commonly seen network solids are compounds of carbon (diamond) and silicon (SiO_2—quartz). The hardness of diamond is due to the network structure formed by carbon atoms. There are no natural seams in diamond's structure in which bonds can be broken.

Metallic Bonds

Think of metallic substances as groups of nuclei surrounded by seas of mobile electrons. Similar to ionic and network substances, a metallic substance can be visualized as one large molecule. Most metals are hard, but the delocalization of electrons makes metals malleable. *Malleable* means that metals can be bent and stretched. All metals except mercury are solids at room temperature, and most metals have high boiling and melting points, though not as high as many ionic and network solids.

Conduction of electricity requires the movements of electrons, and the electrons in a metallic substance can move freely throughout the substance. Because of the sea of freely moving electrons, metals are very good conductors of heat and electricity.

Dipole-Dipole Forces

Dipole-dipole forces occur between neutral, polar molecules: The positive end of one polar molecule is attracted to the negative end of another polar molecule.

Molecules with greater polarity will have greater dipole-dipole attraction. However, dipole-dipole attractions are relatively weak, and these substances melt and boil at very low temperatures. Most substances held together by dipole-dipole attraction are gases or liquids at room temperature.

Hydrogen Bonding

A hydrogen bond is a special kind of dipole-dipole interaction. In a hydrogen bond, the positively charged hydrogen end of a molecule is attracted to the negatively charged end of another molecule containing an extremely electronegative element (fluorine, oxygen, or nitrogen—F, O, N).

Hydrogen bonds occur between a hydrogen atom attached to an electronegative atom and the lone pair of another electronegative atom. In a hydrogen bond, the H atom is shared between the two electronegative atoms. Look at the individual molecules that experience hydrogen bonding: What type of bond does HF have? It's a polar covalent bond. The electron pair shared by H and F spends most of its time on F and this results in a partial positive charge on H and a partial negative charge on F.

$$\text{dipole} \rightarrow \delta^+ \quad \delta^- \leftarrow \text{dipole}$$
$$\text{H} - \text{F}$$

What happens when an HF sees another HF molecule? The electron poor H and the electron rich F are attracted to each other because they can complement each other's partial charges. The lone pair of electrons on F and the partially positive H atom come together and form a partial bond, a hydrogen bond.

$$\overset{\delta^+}{\text{H}} - \overset{\delta^-}{\text{F}} \cdots \overset{\delta^+}{\text{H}} - \overset{\delta^-}{\text{F}}$$

with "hydrogen bond" labeling the dotted line and "covalent bond" labeling the H—F bond.

Hydrogen bonds can occur *only* between an H atom bonded to F, O, or N, and another F, O, or N atom. Some examples of other substances that have hydrogen bonds are water and ammonia. Compounds that have hydrogen bonds typically have higher melting and boiling points than substances that are held together by dipole-dipole forces.

Hydrogen bonding makes water a unique compound with unusual properties. For example, water is unique in that it is less dense as a solid than as a liquid because of its hydrogen bonds. When water forms a solid, the shape of its hydrogen bonds forces the molecules in ice to form a crystal structure, which keeps them farther apart than they are in liquid form. That's why ice floats on water.

Lesson 8: Phase Changes

LONDON DISPERSION FORCES

London dispersion forces occur between neutral, nonpolar molecules. These very weak attractions occur because of the random motions of electrons on atoms within molecules. At a given moment, a nonpolar molecule might have more electrons on one side than the other, giving it an instantaneous polarity. For that fleeting instant, the molecule will act as a very weak dipole.

London dispersion forces are even weaker than dipole-dipole forces, so substances that experience only London dispersion forces melt and boil at low temperatures and tend to be gases at room temperature.

weak attraction between nonpolar molecules

|–| ·····|–|

NAMING THE PHASE CHANGES

Solid to liquid	Melting
Liquid to solid	Freezing
Liquid to gas	Vaporization
Gas to liquid	Condensation
Solid to gas	Sublimation
Gas to solid	Deposition

In a solid or liquid, some molecules at the surface will have enough energy to escape into the gas phase. The pressure exerted by these molecules is called the **vapor pressure** of the substance. As the temperature increases, more and more molecules will have enough energy to escape into the gas phase and the vapor pressure of a liquid will increase. When the vapor pressure of a liquid increases to the point in which it is equal to the surrounding atmospheric pressure, the liquid boils (vaporizes).

THE PHASE DIAGRAM

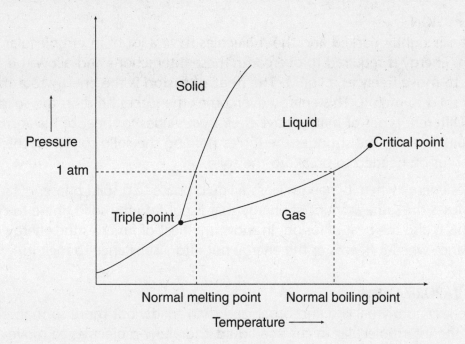

Phase diagrams show the phase (solid, liquid, or gas) of a substance at different temperature and pressure conditions. The diagram above shows that at low temperatures and high pressure the substance will be in a solid state. This makes sense intuitively because at high pressure the molecules of the substance will be pushed close together and at low temperatures the molecules have little kinetic energy. The behaviors of molecules at these conditions are consistent with what you know about solids.

Phase diagrams are divided into three areas, each representing a different phase: solid, liquid, and gas. The lines that separate the phases show temperature and pressure conditions in which two phases can coexist. At these points, there is **equilibrium** between two phases. The **triple point** shows the temperature and pressure where all three phases can exist in equilibrium. The **critical point** is the temperature and pressure condition beyond which the substance is no longer a gas or a liquid. The molecules have too much kinetic energy to remain in liquid form, but are too densely packed to behave as a gas. The substance becomes a special phase called **plasma**.

At 1 atm, normal atmospheric pressure, the temperature at which a substance can coexist as a solid and liquid is the normal melting point and the temperature at which a substance can coexist as liquid and gas is the normal boiling point.

HEATS OF FUSION AND VAPORIZATION

HEAT OF FUSION

Solid matter is tightly packed and the molecules have a lot of intermolecular interaction. Energy is required to overcome these interactions and allow the molecules to move freely as a liquid. The **heat of fusion** is the energy that must be put into a solid to melt it. This energy overcomes the forces holding the solid together. Different types of matter have their own values for heat of fusion. This makes intuitive sense, the stronger the forces holding the solid together, the greater the heat of fusion needed to break up the solid.

Matter loses energy when it changes from a liquid phase to a solid phase. In fact, the substance loses the same amount of energy that is put into the solid phase to cause it to melt. This is also the heat of fusion. In short, the heat of fusion is the energy given off by a substance when it freezes or the energy put into a substance to melt it.

HEAT OF VAPORIZATION

Liquids are less tightly packed and organized than solids, but more so than gases. To overcome the intermolecular forces in a liquid and allow molecules to move as a gas, energy must be added. The **heat of vaporization** is the energy that must be put into a liquid to vaporize it into a gas. Going from a gas to a liquid, the same amount of heat is given off by a substance when it condenses into a liquid. That is, energy must be put in to break up intermolecular interactions when a liquid vaporizes, and energy is released when a gas condenses and gains intermolecular interactions. It makes sense, then, that the stronger the intermolecular forces that hold the liquid together, the greater the heat of vaporization will be.

SPECIFIC HEAT CAPACITY

Matter has the ability to absorb heat, but the amount of heat each substance can absorb varies. As a substance absorbs heat, its temperature increases, and this turns out to be a convenient way to compare different types of matter. Specific heat describes how much heat per mass of substance can be absorbed. **Specific heat** is the amount of heat required to raise the temperature of one gram of a substance one degree Celsius. In other words, the specific heat tells us how much the temperature of a substance changes when you add a certain amount of heat to it. The higher the value of a substance's specific heat, the harder it is to change the substance's temperature. Water has a very large value for specific heat, so it takes more heat to change the temperature of water than most other substances. You can use the formula below to solve problems involving specific heats.

$$q = mc \Delta T$$

q = heat absorbed (J or cal)

m = mass of the substance (g)

c = specific heat capacity

ΔT = temperature change (K or °C)

If you have 4.1g of water at 20°C, what will be the new temperature after 120 J of heat is added? The specific heat of water is 4.18 J/g-°C.

Solve for ΔT.

$$\Delta T = \frac{q}{mc}$$

Now put in the values.

$$\Delta T = \frac{120 \text{ J}}{(4.1 \text{ g})(4.18 \text{ J/g-°C})} = 7.0°C$$

The water temperature increases by 7°C because you have added heat. The new temperature of the water will be 27°C.

Lesson 8: Phase Changes

THE HEATING CURVE

A heating curve shows the change in temperature of a substance as heat is added and the substance undergoes phase changes. A heating curve for water is shown below. Fusion is another term for melting.

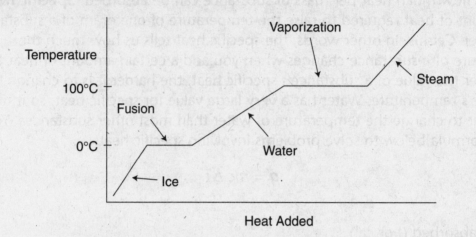

Watch what happens as you follow the curve from the lower left to the upper right. You start out with ice at a low temperature. As you add heat, the temperature of the ice increases until it hits 0°C. That's when ice melts to water. At this point, the curve flattens out because the heat you are adding is used to break the solid bonds instead of increasing the temperature.

The curve stays flat as you keep adding heat until all of the intermolecular interactions are weakened enough for melting to occur. Once the ice is melted, any added heat raises the temperature of liquid water. The temperature continues to increase as you add heat until you reach 100°C, the boiling point of water.

At 100°C, the water begins to vaporize (boil), and the curve flattens out again because the heat is used to break bonds. The curve stays flat until all of the liquid bonds are broken and all of the water has been turned into steam. Once all of the water is gone, the temperature of the steam can begin to increase as you add more heat.

REVIEW FOR LESSON 8

1. Which of the following intermolecular forces is exhibited by water molecules?
   - A ionic bonding
   - B network bonding
   - C hydrogen bonding
   - D metallic bonding

2. Which of the following substances is a solid at room temperature?
   - F MgO
   - G CO_2
   - H H_2O
   - J NO

3. Metals are good conductors of electricity because—
   - A they are solids at room temperature
   - B they are hard, shiny, and malleable
   - C they exhibit ionic bonding when they combine with nonmetals
   - D they have electrons that are free to move around

Questions 4–6 are based on the diagram below.

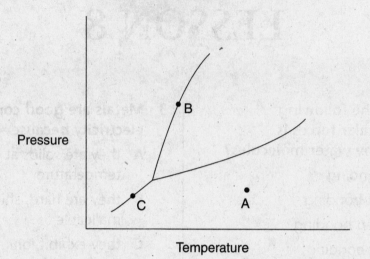

4 Which phase is represented by Point A?

F solid
G liquid
H gas
J plasma

5 What is the name of a phase change that could be taking place at Point B?

A freezing
B sublimation
C condensation
D boiling

6 Which two phases can exist in equilibrium at Point C?

F gas and liquid
G gas and solid
H liquid and solid
J liquid and plasma

7 The heat of vaporization of a substance is—

A equal to the boiling point
B equal to the freezing point
C the heat required to boil the substance
D the heat required to melt the substance

8 The specific heat capacity of aluminum is 0.900 J/g-°C. How much heat is required to raise the temperature of a 55.0-gram piece of aluminum by 10.0°C?

F 4.95 J

G 6.11 J

H 495 J

J 611 J

9 The specific heats of four metals are listed in the table below.

Metal	Specific Heat (J/g-°C)
Au	0.13
Cu	0.39
Fe	0.44
Hg	0.14

If the same amount of heat is added to each of the metals listed in the table, which one will undergo the smallest change in temperature? (The metal samples all have the same mass.)

A Au

B Cu

C Fe

D Hg

10 Ethanol has a specific heat of 2.5 J/g-°C. If 150 J of heat are added to 25 grams of ethanol, what will be the change in temperature of the ethanol?

F 0.4°C

G 2.4°C

H 15°C

J 9400°C

Review for Lesson 8

11 The heat of fusion for water is 6 kJ/mole. If a 2-mole sample of ice has been heated to its melting point, how much additional heat must be added to melt the sample?

A 3 kJ
B 4 kJ
C 8 kJ
D 12 kJ

12 The heat of vaporization of water is 40 kJ/mole. If a 180-gram sample of water has been heated to its boiling point, how much additional heat must be added to boil the sample?

F 4.5 kJ
G 220 kJ
H 400 kJ
J 7200 kJ

ANSWERS AND EXPLANATIONS

1. **C is correct.** Water (H_2O) exhibits hydrogen bonding, where a hydrogen atom on one water molecule is attracted to an oxygen molecule on another water molecule.

2. **F is correct.** MgO contains a metal and a nonmetal, so it exhibits ionic bonding. Substances with ionic bonds are solids at room temperature. Choice **G**, CO_2, exhibits London dispersion forces. Choice **H**, H_2O, exhibits hydrogen bonding. Choice **J**, NO, exhibits dipole-dipole forces. You should know from personal experience that CO_2 and H_2O are not solids at room temperature, so you can eliminate choices **G** and **H** by POE.

3. **D is correct.** Electricity is conducted by the movement of free electrons. In metallic bonding, electrons are free to move throughout the solid, so metals are good conductors of electricity.

4. **H is correct.** Point A is in the region of high temperature and low pressure, which is where gases exist.

5. **A is correct.** Point B is on the line between liquid and solid phase, so the phase change occurring could be either freezing or melting. Freezing is the only correct choice listed.

6. **G is correct.** Point C is on the line between gas and solid phases. By the way, the phase change between gas and solid can be either sublimation or deposition.

7. **C is correct.** The heat of vaporization is the heat required to break the bonds in a liquid and convert the substance to a gas. That's boiling.

8. **H is correct.** Use the formula for specific heat capacity.

$$Q = mc \Delta T$$

$$Q = (55.0 \text{ g})(0.900 \text{ J/g-°C})(10.0°C) = 495 \text{ J}$$

9. **C is correct.** The higher the specific heat, the smaller the temperature change will be for a given amount of heat. Fe has the highest specific heat listed on the table.

10. **G is correct.** Start with the formula for specific heat capacity.

$$Q = mc\,\Delta T$$

Then solve the formula for ΔT.

$$\Delta T = \frac{Q}{mc}$$

Now put in the values.

$$\Delta T = \frac{(150\text{ J})}{(25\text{ g})(2.5\text{ J/g-°C})} = 2.4\text{ C}$$

11 D is correct. If it takes 6 kJ of heat to melt 1 mole of ice, then it must take 12 kJ of heat to melt 2 moles.

12 H is correct. First we have to convert from grams of water to moles of water.

$$\text{Moles} = \frac{\text{grams}}{\text{molecular weight}}$$

$$\text{Moles} = \frac{180}{18}\text{ moles} = 10\text{ moles of water}$$

If it takes 40 kJ of heat to boil 1 mole of water, then it takes 400 kJ of heat to boil 10 moles.

LESSON 9
SOLUTIONS

VOCABULARY

Mixtures of matter can be divided into two types: **homogeneous** and **heterogeneous.** Homogeneous mixtures are those that mix together completely and can't be easily separated. Heterogeneous mixtures are those in which the different types of matter *don't* completely mix together. For example, when sugar dissolves in ice tea, a homogeneous mixture is formed. The two substances are completely mixed together down to the molecular level. On the other hand, when oil is mixed with water, a heterogeneous mixture is formed. The molecules involved stick to their own kind. **Solutions** are homogeneous mixtures where one substance dissolves into another. In a solution, the substance present in the larger proportion is called the **solvent**. The substance present in the smaller proportion is called the **solute**.

Solutions can involve any of the three phases of matter. For instance, you can have a solution of two gases, a solution of a solid in a liquid, or any other combination. However, most of the solutions that you'll see have a liquid as the solvent. Those are the solutions that you're used to seeing, such as salt water, which has solid salt dissolved in water, seltzer water, which has carbon dioxide gas dissolved in water, or vinegar, which has liquid acetic acid dissolved in water.

A quick look around you will demonstrate that most of the solutions that you usually see have water as the solvent. Lemonade, ocean water, tea, soda, and corn syrup are some examples. When a solution has water as the solvent, it is called an **aqueous** solution.

The concentration of a solution tells you how much solute is dissolved in the solvent. A solution that is **concentrated** has a greater proportion of solute than a solution that is **diluted**. A solution that is **saturated** is holding as much solute as it can possibly hold in the solvent. If any more solute is added to a saturated solution, a solid, called a **precipitate**, will form on the bottom of the container. A solution that is unsaturated can hold more solute than it is currently holding. Sometimes, by carefully manipulating the solution, you can get it to hold more solute than it normally would when it is saturated. This is called a **supersaturated** solution.

CONCENTRATION MEASUREMENTS

MOLARITY

Molarity (M) expresses the concentration of a solution in terms of volume. It is the most widely used unit of concentration, turning up in calculations involving equilibrium, acids and bases, and electrochemistry among other things.

When you see a chemical symbol in brackets, that means that they are talking about molarity. For instance "[Na$^+$]" is the same as "the molar concentration (molarity) of sodium ions."

$$\text{Molarity } (M) = \frac{\text{moles of solute}}{\text{liters of solution}}$$

You should know how to find molarity by converting some mass of a substance into moles, then find the concentration in some volume of water. Look at an example of a typical problem: If 7.3 grams of HCl are mixed with water to form 2 liters of solution, what is the concentration of HCl?

First, rephrase this in your own words: The question is asking for the concentration, which means moles per liter. You know the liters involved, but how many moles of HCl do you have? You're given the grams of HCl, so you're in business because you know how to convert grams into moles. First, find the number of moles of HCl by looking up the molecular weight of H and Cl on the periodic table.

$$\text{Moles} = \frac{\text{grams}}{\text{molecular weight}}$$

$$\text{Moles of HCl} = \frac{7.3 \text{ grams}}{36.5 \text{ grams / mole}} = 0.2 \text{ moles of HCl}$$

Now calculate the concentration.

$$\text{Molarity} = \frac{\text{moles of solute}}{\text{liters of solution}}$$

$$[\text{HCl}] = \frac{0.2 \text{ moles}}{2 \text{ liters}} = 0.1 \text{ } M$$

MOLALITY

Molality (m) expresses concentration in terms of the mass of solvent. It's the unit of concentration used for determining the effect of most colligative properties, in which the number of moles of solute is more important than the nature of the solute. Notice that molality is symbolized with a lowercase "m," while molarity is symbolized with an uppercase "M."

$$\text{Molality } (m) = \frac{\text{moles of solute}}{\text{kilograms of solvent}}$$

If 21 grams of NaF are dissolved in 0.25 kg of water, what is the molality of the solution?

First find the number of moles of NaF. Look up the molecular weight of Na and F on the periodic table.

$$\text{Moles} = \frac{\text{grams}}{\text{molecular weight}}$$

$$\text{Moles of NaF} = \frac{21 \text{ grams}}{42 \text{ grams/mole}} = 0.5 \text{ moles of NaF}$$

Now calculate the molality.

$$\text{Molality} = \frac{\text{moles of solute}}{\text{kg of solvent}}$$

$$\text{Molality of NaF} = \frac{0.5 \text{ moles}}{0.25 \text{ kg}} = 2.0 \, m \text{ NaF}$$

Molarity and molality differ in two ways:

- Molarity tells you about moles of solute per *volume* (L) of the *entire solution* (that is, the solute and the solvent).

- Molality tells you about moles of solute per *mass* (kg) of the *solvent*.

Lesson 9: Solutions

MOLE FRACTION

Mole fraction gives the fraction of moles of a given substance out of the total moles present in a sample. It is used in determining how the vapor pressure of a solution is lowered by the addition of a solute.

$$\text{Mole fraction} = \frac{\text{moles of substance}}{\text{total number of moles in solution}}$$

A solution contains 35 moles of water and 15 moles of ethanol. What is the mole fraction of ethanol in the solution?

$$\text{Mole fraction of ethanol} = \frac{\text{moles of ethanol}}{\text{total number of moles in solution}}$$

$$\text{Mole fraction of ethanol} = \frac{15}{15 + 35} = \frac{15}{50} = 30\%$$

NORMALITY

Some solutes dissolve and break up into ions, such as salts, acids, and bases. In special cases you may be interested in something other than the actual number of moles of solute in the solution. For instance, when dealing with acids, you may be interested in the concentrations of H^+ ions, no matter where they came from. If you put 1 mole of H_2SO_4 in water, you get 2 moles of H^+ ions. So what happens if you make a 1 M solution of H_2SO_4? You get a 2 M solution of H^+ ions and you need a number to describe this fact. You use **normality** for that.

1 mole of H_2SO_4 produces 2 **equivalents** of H^+.

$$\text{Normality } (N) = \frac{\text{equivalents}}{\text{liters of solution}}$$

SOLUTES AND SOLVENTS

There is a basic rule for remembering what solutes will dissolve in what solvents:

Like dissolves like.

That means that polar or ionic solutes (like salt) will dissolve in polar solvents (like water). That also means that nonpolar solutes, like organic compounds, are best dissolved in nonpolar solvents.

When ionic substances dissolve, they **dissociate** into ions. Free ions in a solution are called **electrolytes**, because the solution can conduct electricity. Some salts dissociate completely into ions and others will partially dissociate, which means that the ions will remain close to each other in pairs rather than being independent and fully surrounded by solvent. Solutes that completely dissociate are called strong electrolytes, and those that remain ion-paired to some extent are called weak electrolytes. Which do you think will produce a solution that better conducts electricity? Solutions of strong electrolytes conduct better than those of weak electrolytes.

Different compounds will dissociate into different numbers of particles. Some won't dissociate at all, and others will break up into several ions. The **van't Hoff factor** (i) tells how many ions one unit of a substance will produce in a solution. For instance:

$C_6H_{12}O_6$ is nonpolar and does not dissociate, so $i = 1$.

NaCl dissociates into Na^+ and Cl^-, so $i = 2$.

HNO_3 dissociates into H^+ and NO_3^-, so $i = 2$.

$CaCl_2$ dissociates into Ca^{2+}, Cl^- and Cl^-, so $i = 3$.

COLLIGATIVE PROPERTIES

When a solute is added to a solvent, the boiling point, freezing point, and vapor pressure of the solution will be different from those of the solvent alone. These changes are caused by **colligative properties**. Colligative properties are properties of a solution that depend on the *number* of solute particles in the solution rather than the *type* of particle. For colligative properties, the identity of the particle is not important. That is, if you have a 1 M solution of any solute, the change in a colligative property will be the same no matter what the size or charge of the solute particles. For example, a 1 M solution of any substance will lower the freezing point of water. That means 1 M glucose, ethanol, or anything else soluble in water. But think about a salt like NaCl: It would take only a 0.5 M solution because the ions fully dissociate to generate 1 M particles total. For a colligative property, it doesn't matter the type or nature of the particle.

BOILING POINT ELEVATION

Think about boiling water: The molecules of water in the liquid phase escape into a gas phase at the surface. When a solute is added to a solution, solute particles take up space at the surface of the solution, and fewer molecules can escape. What happens to the boiling point? In order for the molecules to escape, they have to have more energy than they did without the solute. As a result, the boiling point increases. How much does the boiling point increase? That depends upon the number of particles *in molality*, the type of solvent, and the number of particles the solute dissolves into (that is, the van't Hoff factor). You can figure out the boiling point elevation of a solution by using the formula below.

$$\Delta T = ik_b m$$

i = the van't Hoff factor, the number of particles into which the added solute dissociates

k_b = the boiling point elevation constant for the solvent

m = molality

Example:

What is the boiling point of a 3 m solution of NaCl in water? The boiling point elevation constant for water is 0.5°C/m.

NaCl breaks up into Na^+ and Cl^-, so i for NaCl is 2.

$$\Delta T = ik_b m$$

$$\Delta T = (2)(0.5°C/m)(3m)°C = 3°C$$

The boiling point elevation is 3°C. Because the normal boiling point of water is 100°C, the boiling point of the NaCl solution is 103°C.

FREEZING POINT DEPRESSION

Solids are held together by attractive intermolecular forces. What happens when you add a solute to a liquid, then try to freeze the solution? The solute particles get in the way of molecules trying to pack together in the freezing process. As a result, it's harder to freeze the solution and the freezing point of the solution decreases. That is, more energy must be removed for the liquid to form a solid. Some good news: The formula for freezing point depression is exactly the same as the formula for boiling point elevation, except that the temperature is going down instead of up.

$$\Delta T = ik_f m$$

i = the van't Hoff factor, the number of particles into which the added solute dissociates

k_f = the freezing point depression constant for the solvent

m = molality

Example:

What is the freezing point of a 1.00*m* aqueous solution of $MgCl_2$? The freezing point depression constant for water is 1.86°C/*m*.

$MgCl_2$ breaks up into Mg^{2+} and 2 Cl^-, that's three particles total, so *i* for $MgCl_2$ is 3.

$$\Delta T = ik_f m$$

$$\Delta T = (3)(1.86°C/m)(1.00\ m) = 5.58°C$$

You have to keep the numbers straight here: The freezing point of water is *not* 5.58°C now. This is the number of degrees the freezing point is depressed, so subtract this from the normal freezing point of water. Because the normal freezing point of water is 0°C, the freezing point of the $MgCl_2$ solution is –5.58°C.

Vapor Pressure Depression

Remember boiling point elevation: If you add solute to boiling water, the particles take up space at the surface where liquid water escapes as a gas. This is the same as saying that the vapor pressure is decreased by the addition of solute. The vapor pressure of a liquid is the pressure of the vapor in equilibrium with the liquid and is dependent upon temperature. What happens as you increase the temperature? The vapor pressure increases, and eventually the vapor pressure will become high enough for the liquid to boil. However, when a solute is added to a solution, the vapor pressure of the solution will decrease. This happens for the same reasons that the boiling point increases: the solute particles get in the way at the surface and make it harder for particles to escape. The formula for vapor pressure lowering uses *mole fractions* instead of molality.

$$P = XP°$$

P = vapor pressure of the solution

$P°$ = vapor pressure of the pure solvent

X = the mole fraction of the solvent

SOLUBILITY

When you dissolve salt into water, you can keep adding salt and it will continue to dissolve until the solution is **saturated**. At this point, when you add salt it just sits on the bottom and no more will dissolve. The **solubility** of a substance describes how much of it can be dissolved in a solution before the solution becomes saturated. Solubility is usually measured in grams of solute per 100 mL of water. Below is a typical chart of solubilities, which shows how the solubility of different salts changes with different temperatures.

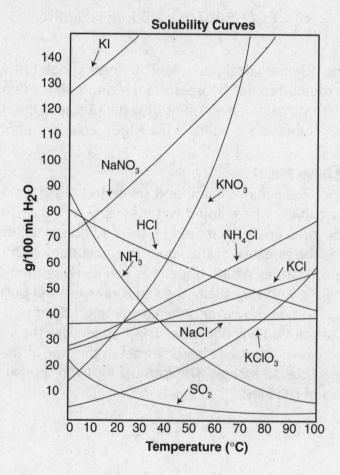

Because solubility of a substance varies with temperature, there isn't just one value to look up. If you are asked to work a solubility problem, you need to know the temperature to find the right solubility constant. Here's how to use this chart: Find the line that represents the salt that you are interested in and look up the solubility for a given temperature. Try finding these for practice: The solubility of $NaNO_3$ at 30°C is roughly 90 g per 100 mL of H_2O. Similarly, the solubility of SO_2 at 50°C is roughly 5 g per 100 mL of H_2O. Generally, the higher up on this chart the line for a salt appears, the more soluble the salt.

REVIEW FOR LESSON 9

1. Which of the following will most likely happen when a solute is added to a saturated solution?

   A The solution will boil.

   B A precipitate will form.

   C The solution will freeze.

   D The solution will become less concentrated.

2. If 30 grams of $CaBr_2$ (molecular weight = 200 g/mole) are dissolved in water to form a 1.5 liter solution, what is the concentration of $CaBr_2$?

   F 0.1 M

   G 0.2 M

   H 0.4 M

   J 0.5 M

3. If 4 moles of KCl are added to 1.6 kg of water, what is the molality of the KCl solution?

   A 0.4 m

   B 1.4 m

   C 2.5 m

   D 4.6 m

4. How many equivalents of H^+ ions are produced by 1 mole of H_2S?

   F 0.5

   G 1

   H 2

   J 4

5 How many moles of MgCl$_2$ are present in 5 liters of a 0.12 M MgCl$_2$ solution?

   A 0.6 moles

   B 1.2 moles

   C 10 moles

   D 30 moles

6 What is the freezing point of a 0.75 m aqueous solution of HCl? The freezing point depression constant for water is 1.86°C/m.

   F 1.4°C

   G 0°C

   H −1.4°C

   J −2.8°C

7 What is the boiling point of a 4.0 m aqueous solution of CaCl$_2$? The boiling point elevation constant for water 0.5°C/m.

   A 100°C

   B 102°C

   C 103°C

   D 106°C

Questions 8–10 are based on the diagram below.

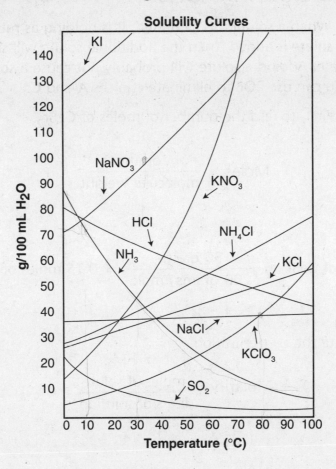

8 What is the maximum amount of NH$_3$ that can be dissolved in 100 mL of water at a temperature of 10°C?

F 10 g
G 30 g
H 70 g
J 100 g

9 Which of the salts listed below is the most soluble at a temperature of 30°C?

A KClO$_3$
B NaNO$_3$
C NaCl
D NH$_4$Cl

10 At what temperature do KClO$_3$ and NaCl have the same solubility?

F 20°C
G 40°C
H 60°C
J 80°C

Review for Lesson 9

ANSWERS AND EXPLANATIONS

1. **B is correct.** When a solution is saturated, it is holding as much solute as it can. If more solute is added, then the additional solute will sink to the bottom as a precipitate. Adding a solute will probably not cause a solution to boil or freeze, so you can use POE to eliminate choices **A** and **C**.

2. **F is correct.** First, to find the number of moles of $CaBr_2$.

$$\text{Moles} = \frac{\text{grams}}{\text{molecular weight}}$$

$$\text{Moles of } CaBr_2 = \frac{30 \text{ grams}}{200 \text{ grams / mole}} = 0.15 \text{ moles of } CaBr_2$$

Now, calculate the concentration.

$$\text{Molarity} = \frac{\text{moles of solute}}{\text{liters of solution}}$$

$$\text{Molarity of } CaBr_2 = \frac{0.15 \text{ moles}}{1.5 \text{ liters}} = 0.1 \text{ } M$$

3. **C is correct.** Use the formula for molality.

$$\text{Molality} = \frac{\text{moles of solute}}{\text{kilograms of solvent}}$$

$$\text{Molality of } KCl = \frac{4 \text{ moles}}{1.6 \text{ kg}} = 2.5 \text{ } m \text{ } KCl$$

4   **H is correct.** 1 mole of H_2S contains 2 moles of H^+ ions, so 1 mole of H_2S will produce 2 equivalents of H^+ ions.

5   **A is correct.** Use the molarity formula and solve for moles.

$$\text{Molarity} = \frac{\text{moles of solute}}{\text{liters of solution}}$$

$$\text{Moles} = (\text{molarity})(\text{liters})$$

$$\text{Moles of } MgCl_2 = (0.12\ M)(5\ L) = 0.6 \text{ moles of } MgCl_2$$

6   **J is correct.** Use the formula for freezing point depression. HCl breaks up into two particles, H^+ and Cl^-, so i is 2.

$$\Delta T = ik_f m$$

$$\Delta T = (2)(1.86°C/m)(0.75\ m) = 2.8°C$$

The freezing point depression is 2.8°C. Because the normal freezing point of water is 0°C, the new freezing point is −2.8°C.

7   **D is correct.** Use the formula for boiling point elevation. $CaCl_2$ breaks up into three particles, 2 Cl^- and Ca^{2+} so i is 3.

$$\Delta T = ik_b m$$

$$\Delta T = (3)(0.5°C/m)(4.0\ m) = 6°C$$

The boiling point elevation is 6°C. Because the normal boiling point of water is 100°C, the new boiling point is 106°C.

8 **H is correct.** The NH_3 line on the diagram is at about 70 g/100 mL H_2O when the temperature is 10°C.

9 **B is correct.** At 30°C, the $NaNO_3$ line is the highest of the salts listed on the solubility graph. That means that it is the most soluble.

10 **J is correct.** The lines for $KClO_3$ and $NaCl$ intersect at about the 80°C mark, so at that point they are equally soluble.

LESSON 10
EQUILIBRIUM

REVERSIBLE REACTIONS
Some chemical reactions are reversible. Although you write an equation as reactants going to products, the products also can react and form the reactants. When a reaction begins with starting material, only the forward reaction occurs, but as product forms, some of it will convert back to starting material. That is, the reactants are becoming products and the products are converting to the reactants at the same time. Eventually, the process will reach a situation where the two reactions will cancel each other's effects and the concentrations of reactants and products will remain constant. This situation is called **dynamic equilibrium**. The forward and reverse reactions are still going on, but the concentrations don't change.

A reaction is at equilibrium when the rate of the forward reaction is equal to the rate of the reverse reaction.

THE EQUILIBRIUM CONSTANT, K_{eq}
Equilibrium is a special situation in which the concentrations of all reactants and products are constant. The relationship between the concentrations of reactants and products at equilibrium is given by the **equilibrium expression**. You can look at any reaction and derive the equilibrium equation yourself. The following steps show how to derive the equilibrium expression from the reaction formula.

For the reaction:

$$aA + bB \rightleftarrows cC + dD$$

$$K_{eq} = \frac{[C]^c[D]^d}{[A]^a[B]^b}$$

1. [A], [B], [C], and [D] are **molar concentrations** or **partial pressures** at equilibrium.

2. Products are in the numerator, and reactants are in the denominator.

3. Coefficients in the balanced equation become exponents in the equilibrium expression.

4. Solids and pure liquids are ignored.

5. Don't worry about the units for K_{eq}.

Example 1 $2 H_2S(g) + 3 O_2(g) \rightleftarrows 2 H_2O(g) + 2 SO_2(g)$

$$K_{eq} = \frac{[H_2O]^2[SO_2]^2}{[H_2S]^2[O_2]^3}$$

Notice how the coefficients in the balanced equation become the exponents.

Example 2 $HC_2H_3O_2(aq) \rightleftarrows H^+(aq) + C_2H_3O_2^-(aq)$

$$K_{eq} = \frac{[H^+][C_2H_3O_2^-]}{[HC_2H_3O_2]}$$

This reaction shows the dissociation of acetic acid in water. All of the reactants and products are aqueous particles, so they are all included in the equilibrium expression. None of the reactants or products have coefficients, so there are no exponents in the equilibrium expression. This is the standard form of K_a, the acid dissociation constant.

Example 3 $CaF_2(s) \rightleftarrows Ca^{2+}(aq) + 2 F^-(aq)$

$$K_{eq} = [Ca^{2+}][F^-]^2$$

This reaction shows the dissociation of a slightly soluble salt. There is no denominator in this equilibrium expression because the reactant is a solid. Solids are left out of the equilibrium expression because if a solid is present, then the solution is saturated with that solute and the concentration of the dissolved solute will not change. That is, the solid won't contribute to the concentration of the dissolved solute. This form of K_{eq} is called the solubility product, K_{sp}.

Example 4 $NH_3(aq) + H_2O(l) \rightleftarrows NH_4^+(aq) + OH^-(aq)$

$$K_{eq} = \frac{[NH_4^+][OH^-]}{[NH_3]}$$

This is the acid-base reaction between ammonia and water. You can leave water out of the equilibrium expression because it is a pure liquid. The concentration of water is so large (about 55 molar) that it will not change significantly during the reaction. You can consider it constant. This is the standard form for K_b, the base dissociation constant.

What does the equilibrium constant really tell us? The K_{eq} tells you the relative amounts of products and reactants when the reaction reaches equilibrium. Think about what the value of K_{eq} means: If it's a large number, it means that the numerator is very large compared to the denominator, which means that at equilibrium, the reaction has more products than reactants. But what if the value for K_{eq} is small? Then the opposite is true: The denominator is large and the numerator is small. This means that the concentration of reactants is greater than the concentration of products at equilibrium. Keep in mind that the K_{eq} value is true only at equilibrium.

An amazing thing about the equilibrium constant is that *it remains the same no matter how you try to change the concentrations or products and reactants*. If you add more of any or all of the reactants and products, these concentrations will adjust themselves until the equilibrium expression is equal to K_{eq}. However, the equilibrium constant is temperature dependent, so it will be affected by changes in temperature. That is, K_{eq} will have different values at different temperatures.

THE REACTION QUOTIENT, Q

When a reaction is not at equilibrium, you will want to find the **reaction quotient**, Q. The reaction quotient can be used to predict the direction in which a reaction will proceed from a given set of initial conditions. The good news is that the reaction quotient is determined in exactly the same way as the equilibrium constant, but initial conditions are used in place of equilibrium conditions.

For the reaction:

$$aA + bB \rightleftarrows cC + dD$$

$$Q = \frac{[C]^c[D]^d}{[A]^a[B]^b}$$

So Q is exactly the same as K_{eq} except that [A], [B], [C], and [D] are initial molar concentrations instead of equilibrium concentrations.

The value of Q is helpful because if you also know the value of K_{eq}, then you can compare them to see whether the reaction will go forward or backward. What will the relative values of Q and K_{eq} tell us? The value of Q will always want to move towards the value of K. That is, the reaction will progress such that Q comes closer to K.

- If Q is *less* than the calculated K for the reaction, the reaction proceeds forward, generating products. That is, Q wants to go toward the value of K, which means the reaction will want to go toward making products.

- If Q is *greater* than K, the reaction proceeds backward, generating reactants. Because Q wants to change to become like K, the reaction will proceed backward toward reactants.

- If Q = K, the reaction is already at equilibrium.

LE CHATELIER'S PRINCIPLE

A chemical equilibrium will resist change. If you alter some aspect of an equilibrium situation, such as the temperature, pressure, or concentration of one of the reactants or products, the reaction will shift to restore equilibrium. This is called **Le Chatelier's Principle,** which says that *whenever a stress is placed on a reaction at equilibrium, the reaction will shift to restore equilibrium by relieving that stress*. For example, if you add more of a reactant, the reaction will proceed toward the products until the concentrations are at equilibrium again.

For the following sections, use the Haber process as an example. The Haber process is used in the industrial preparation of ammonia.

$$N_2(g) + 3\,H_2(g) \rightleftarrows 2\,NH_3(g) + \text{Heat}$$

CONCENTRATIONS

When the concentration of a reactant or product is increased, the reaction will proceed in the direction that will use up the added substance.

What happens if N_2 is added? The reactant side gets crowded, and the reaction proceeds in the forward direction to consume the added N_2 until equilibrium is reached. You can think of the added N_2 as a stress that disturbs the equilibrium, and to ease the stress, the reaction consumes the added material to restore equilibrium. You can also consider the effect on Q and think about what direction Q would have to go to reach K_{eq}. Now what does the reaction do if NH_3 is added? The products are too crowded and the reaction goes in the reverse direction in order to reestablish equilibrium.

The reverse is also true.

When the concentration of a reactant or product is decreased, the reaction will proceed in the direction that will produce more of the substance that has been removed.

For example, if N_2 or H_2 is removed, it creates space on the reactant side, so the reaction will proceed in the reverse direction to fill the space. If NH_3 is removed, space is created on the product side, and the reaction will proceed in the forward direction to make NH_3.

PRESSURE

When the pressure at which a reaction takes place is increased, the reaction will proceed in the direction that produces fewer moles of gas.

When the pressure for a reaction is increased, the gas molecules get crowded, and the reaction will shift to try to relieve the crowding. The Haber process proceeds in the forward direction because the products have fewer moles of gas (2) than the reactants (4).

When the pressure at which a reaction takes place is decreased, the reaction will proceed in the direction that produces more moles of gas.

When the pressure for the Haber process is decreased, more space for gas is made available, and the reaction will shift to fill that space. The reactants have more moles of gas (4) than the products (2), so the reaction will shift to favor the reverse reaction.

If there is no gas involved in the reaction, or if the reactants and products have the same number of moles of gas, then changes in pressure have no effect on the equilibrium.

Temperature

*When temperature is increased, the reaction will proceed in the **endo**thermic direction.*

You can think of heat as if it were just like any other reactant or product in a reaction. In the endothermic direction, heat is used up, so it is a reactant. In the exothermic direction, heat is produced, so it is a product. When the temperature for the Haber process is increased, it's the same as adding heat, so the product side gets crowded, and the reaction will shift to favor the reactants. So when heat is added, the endothermic direction is favored.

*When temperature is decreased, the reaction will proceed in the **exo**thermic direction.*

When the temperature is decreased, it's the same as removing heat. That creates a space on the product side of the Haber process, so the reaction will shift to favor the creation of products. So when the temperature for the Haber process is decreased, the reaction proceeds in the forward direction because the forward reaction is exothermic.

REVIEW FOR LESSON 10

1. A reversible reaction is at equilibrium when—
   A the concentration of products is equal to the concentration of reactants
   B there are no more reactants remaining in the reaction chamber
   C the rate of the reverse reaction is equal to the rate of the forward reaction
   D the concentration of products exceeds the concentration of reactants

2. Which of the following is the correct equilibrium expression for the reaction shown below?

   $$2\ NO(g) + O_2(g) \rightleftarrows 2\ NO_2(g)$$

   F $\dfrac{[NO_2]^2}{[NO]^2[O_2]}$

   G $\dfrac{[NO_2]}{[NO][O_2]}$

   H $\dfrac{[NO][O_2]}{[NO_2]}$

   J $\dfrac{[NO]^2[O_2]}{[NO_2]}$

3. When solid AgCl is placed in water, some of the salt will dissociate into Ag^+ and Cl^- ions. Eventually the solid will come to equilibrium with its dissolved ions. Which of the following is the correct equilibrium expression, K_{sp}, for this reversible reaction.

   A $K_{sp} = [Ag^+] + [Cl^-]$

   B $K_{sp} = [Ag^+][Cl^-]$

   C $K_{sp} = [Ag^+][Cl^-][AgCl]$

   D $K_{sp} = \dfrac{[Ag^+]}{[Cl^-]}$

Review for Lesson 10

4 The following reaction took place in a sealed chamber and the equilibrium concentrations were measured.

$$H_2(g) + I_2(g) \rightleftarrows 2\,HI(g)$$

$[H_2] = 0.5\ M$

$[I_2] = 4\ M$

$[HI] = 10\ M$

Using the information given above, calculate the equilibrium constant for this reaction.

F $K_{eq} = 5$
G $K_{eq} = 20$
H $K_{eq} = 50$
J $K_{eq} = 80$

5 For a certain reversible reaction, $K_{eq} = 20$. If the reaction quotient is calculated for a given set of initial concentrations and found to be equal to 10, which of the following statements is true?

A The reaction will proceed in the forward direction to make more products.
B The reaction is initially at equilibrium.
C The reaction will proceed in the reverse direction to make more reactants.
D The reaction will not occur.

6 The solubility products for some salts are given in the chart below.

Salt	Solubility product, K_{sp}
CdS	8.0×10^{-28}
FeS	6.0×10^{-19}
NiS	1.4×10^{-24}
ZnS	3.0×10^{-23}

Which of the salts listed in the chart is the most soluble?

F CdS
G FeS
H NiS
J ZnS

7. Which of the following is the equilibrium expression, K_a, for the acid dissociation reaction shown below?

$$HC_2H_3O_2(aq) \rightleftarrows H^+(aq) + C_2H_3O_2^-(aq)$$

A $\dfrac{[H^+]+[C_2H_3O_2^-]}{[HC_2H_3O_2]}$

B $\dfrac{[H^+]-[C_2H_3O_2^-]}{[HC_2H_3O_2]}$

C $[H^+][C_2H_3O_2^-][HC_2H_3O_2^-]$

D $\dfrac{[H^+][C_2H_3O_2^-]}{[HC_2H_3O_2]}$

Questions 8 and 9 refer to the reversible reaction shown below. Remember that the forward reaction is endothermic.

$$PCl_5(g) + \text{heat} \rightleftarrows PCl_3(g) + Cl_2(g)$$

8. Which of the following will cause an increase in the equilibrium concentration of Cl_2 gas?

F The addition of PCl_3 gas
G The removal of PCl_5 gas
H A decrease in temperature
J An increase in temperature

9. If the volume of the reaction chamber in which the reaction above takes place is decreased, which of the following changes will occur?

A The concentration of PCl_5 will increase.
B The concentration of PCl_3 will increase.
C The concentration of Cl_2 will increase.
D No changes in concentration will occur.

Questions 10 through 12 refer to the reversible reaction shown below.

$$CH_4(g) + H_2O(g) \rightleftarrows CO(g) + 3H_2(g)$$

10. Which of the following changes will occur when CO gas is added to the reaction chamber?

   F There will be an increase in the concentration of CH_4.

   G There will be an increase in the concentration of H_2.

   H There will be a decrease in the concentration of CH_4.

   J There will be a decrease in the concentration of H_2O.

11. Which of the following changes to the equilibrium system will bring about an increase in the concentration of CO gas?

   A The addition of H_2 gas to the system.

   B The removal of H_2O gas from the system.

   C An increase in the volume of the system.

   D A decrease in the volume of the system.

12. If H_2O is removed from the system at equilibrium, which of the following changes will occur?

   F $[CH_4]$ will increase and $[H_2]$ will increase.

   G $[CH_4]$ will decrease and $[H_2]$ will increase.

   H $[CH_4]$ will increase and $[H_2]$ will decrease.

   J $[CH_4]$ will decrease and $[H_2]$ will decrease.

Questions 13 and 14 refer to the reversible reaction shown below. The reaction is endothermic.

$$N_2(g) + O_2(g) + \text{heat} \rightleftarrows 2\,NO(g)$$

13 An increase in the temperature of the reaction chamber will cause—

A an increase in the equilibrium concentration of N_2

B an increase in the equilibrium concentration of O_2

C a decrease in the equilibrium concentration of N_2

D a decrease in the equilibrium concentration of NO

14 Which of the following changes to the equilibrium conditions will not cause a shift in favor of either side of the reaction?

F An increase in temperature

G A decrease in the volume of the reaction chamber

H The addition of N_2 gas

J The removal of O_2 gas

Review for Lesson 10

ANSWERS AND EXPLANATIONS

1. **C is correct.** Chemical equilibrium is defined as the point when the rates of the forward and reverse reactions are equal.

2. **F is correct.** The equilibrium constant places the concentrations of products in the numerator and the concentrations of reactants in the denominator. Also, the coefficients of NO and NO_2 become exponents.

$$K_{eq} = \frac{[NO_2]^2}{[NO]^2[O_2]}$$

If you only remember that products go over reactants, you can still use POE to eliminate two answer choices, **H** and **J**. Similarly, if you only remember that coefficients become exponents, but you can't remember whether reactants or products go on top, you can use POE to eliminate two answer choices, **G** and **H**.

3. **B is correct.** The reversible reaction described is shown below.

$$AgCl(s) \rightleftarrows Ag^+(aq) + Cl^-(aq)$$

Because the reactants of this reaction are solid and can be left out of the equilibrium expression, K_{eq} for this reaction takes the form of the solubility product, K_{sp}, with no denominator.

$$K_{sp} = [Ag^+][Cl^-]$$

4. **H is correct.** First set up the equilibrium expression.

$$K_{eq} = \frac{[HI]^2}{[H_2][I_2]}$$

Now plug in the numbers.

$$K_{eq} = \frac{(10)^2}{(0.5)(4)} = \frac{100}{2} = 50$$

5   A is **correct**. When Q is less than K_{eq}, the reaction proceeds in the forward direction.

6   G is **correct**. The solubility product is the product of concentrations of particles in solution, so the larger the K_{sp}, the more particles in solution and the more soluble the salt. FeS has the solubility product with the least negative exponent, which makes it the largest one listed. If you are guessing, it's probably best to guess either the largest or the smallest number in a problem like this.

7   D is **correct**. The acid dissociation constant, K_a, is just another name for the equilibrium constant, so you make it the same way you would K_{eq} putting the concentrations of products in the numerator and the concentrations of reactants in the denominator.

$$K_a = \frac{[H^+][C_2H_3O_2]}{[HC_2H_3O_2]}$$

8   J is **correct**. Heat is a reactant in this reaction. (The forward reaction is endothermic.) Increasing the temperature means adding heat, which will drive the reaction to produce more products, which will increase the equilibrium concentration of Cl_2 gas. All of the other choices will cause the reaction to go to the left, producing more reactants.

9   A is **correct**. A decrease in volume will cause the reaction to shift toward the side that has fewer moles of gas in the balanced equation. In this reaction, the left side has 1 mole of gas and the right side has 2 moles of gas, one PCl_3 + one Cl_2, so the reaction will shift toward the left, which will increase the concentration of PCl_5 gas. It's best to do this one by POE, crossing off wrong answers as you go.

10   F is **correct**. The addition of CO gas causes crowding on the right side, which will cause the reaction to shift to the left, creating more reactants. This will increase the concentration of CH_4. It's best to do this one by POE, crossing off wrong answers as you go.

11   C is **correct**. If the volume of the system is increased, the reaction will shift to produce more moles of gas to fill the space. There are 2 moles of gas on the left side (one CH_4 + one H_2O) and 4 moles of gas on the right side (one CO and $3H_2$) so increasing the volume will cause the reaction to shift to the right, increasing the concentration of CO. It's best to do this one by POE, crossing off wrong answers as you go.

Review for Lesson 10

12. **H is correct.** The removal of H_2O creates space on the left side of the reaction, and the reaction will shift to try to fill that space. When the reaction shifts to the left, reactants will be created and products will be used up, so the concentration of CH_4 will increase, and the concentration of H_2 will decrease. It's best to do this one by POE, crossing off wrong answers as you go.

13. **C is correct.** Heat is a reactant in this reaction. (The forward reaction is endothermic.) Increasing the temperature means adding heat, which will drive the reaction to produce more products and use up the reactants, which will decrease the equilibrium concentration of N_2 gas. It's best to do this one by POE, crossing off wrong answers as you go.

14. **G is correct.** There are 2 moles of gas on the left side (one N_2 and one O_2) 2 moles of gas on the right side (2 NO). Because there are the same number of moles of gas in the reactants and products, changes in volume will not cause the equilibrium to shift in either direction. It's best to do this one by POE, crossing off wrong answers as you go.

LESSON 11
ACIDS AND BASES

PROPERTIES

Acids and bases are everywhere. They provide crucial roles in chemistry, biology, and even in gastronomy (a.k.a., cooking). You can find all sorts of acids and bases in common household products: lemon, vinegar, and even your own stomach juices are all highly acidic. Many household chemicals, such as ammonia and drain cleaners (aqueous sodium hydroxide), are bases. Acids and bases are also everywhere in chemical problems, so you'll become familiar with the basic (no pun intended) facts before we move on.

ACIDS

- An **acidic solution** is a solution that contains more hydrogen ions (H^+) than hydroxide ions (OH^-). That is, $[H^+]$ is greater than $[OH^-]$.

- Acidic solutions are **electrolytic**. What do you know about electrolytes? They conduct electricity. When an acid is placed into solution, it dissociates into H^+ and the conjugate base anion, and solutions of ions conduct electricity.

- Acidic solutions cause indicators to change colors. **Indicators** are special chemicals that change color when placed in an acidic solution. Indicators are useful to tell whether a solution is acidic or not.

- Acids react with certain metals to form hydrogen gas. After the acid dissociates into H^+ and the conjugate base, the metal gives up electrons to the H^+ to form H_2 gas. What type of reaction is this? A **redox reaction**: H^+ is reduced to H_2 and the metal is oxidized to a cation. Of course, the more reactive a metal is, the more likely it is to react with an acid.

- Acidic solutions have a sour taste. Of course, you're *never* supposed to taste a chemical to see if it is an acid or not, but many common foods are acidic. For example, lemon juice and vinegar are acidic.

Acids react with bases to form water and a salt. That's called a **neutralization reaction**.

BASES

- A basic solution is a solution that contains more hydroxide ions (OH^-) than hydrogen ions (H^+). That is, $[OH^-]$ is greater than $[H^+]$.

- Basic solutions are electrolytic. Like acids, bases conduct electricity because when a base is placed into solution, it breaks up into positive and negative ions, which are then free to move throughout the solution.

- Basic solutions cause indicators to change colors. Indicators have one color when placed in basic solutions and another color when placed in acidic solutions.

- Basic solutions are slippery. A typical example of a basic solution is soapy water.

- Bases react with acids in neutralization reactions to form water and a salt.

DIFFERENT TYPES OF ACIDS AND BASES

Although the main difference between an acid and a base is that an acid has a lot of H^+ ions and a base has a lot of OH^- ions, different scientists have come up with more subtle definitions of acids and bases.

ARRHENIUS

Arrhenius provided the most straight forward definition of an acid and a base: An acid is a substance that ionizes in water and produces hydrogen ions (H^+ ions). For instance, HCl is an acid.

$$HCl \rightarrow H^+ + Cl^-$$

He defined a base as a substance that ionizes in water and produces hydroxide ions (OH^- ions). For instance, NaOH is a base:

$$NaOH \rightarrow Na^+ + OH^-$$

BRONSTED-LOWRY

Bronsted and Lowry defined an acid as a substance that is capable of donating a proton, which is the same as donating an H^+ ion, so there's nothing new here. But their definition of a base is a little more subtle: A Bronsted-Lowry base is a substance that is capable of accepting a proton.

Look at the reversible reaction below.

$$HC_2H_3O_2 + H_2O \leftrightarrow C_2H_3O_2^- + H_3O^+$$

According to Bronsted-Lowry:

$HC_2H_3O_2$ and H_3O^+ are acids.

$C_2H_3O_2^-$ and H_2O are bases.

Now look at this reversible reaction:

$$NH_3 + H_2O \leftrightarrow NH_4^+ + OH^-$$

According to Bronsted-Lowry:

NH_3 and OH^- are bases.

H_2O and NH_4^+ are acids.

So, in each case, the compound with the H^+ ion is the acid and the same compound without the H^+ ion is the base; the two species are called a **conjugate pair**. These are the acid-base conjugate pairs in the above reactions:

Acid	Conjugate Base
$HC_2H_3O_2$	$C_2H_3O_2^-$
NH_4^+	NH_3
H_3O^+	H_2O
H_2O	OH^-

Notice that water can act as either an acid or a base.

LEWIS

Lewis focused on electrons, and his definitions are the most broad of the acid-base definitions. Lewis defined a base as an electron pair donor and an acid as an electron pair acceptor. According to Lewis's rule, all of the Bronsted-Lowry bases above are also Lewis bases and all of the Bronsted-Lowry acids are Lewis acids.

pH AND pOH

The acidity of a solution is usually measured by its pH. The pH scale goes from 0 to 14, and you can tell by looking at the pH whether a solution is acidic or basic.

- If the pH is *less* than 7, the solution is acidic.

- If the pH is *greater* than 7, the solution is basic.

- If the pH is *equal* to 7, the solution is neutral, like pure water.

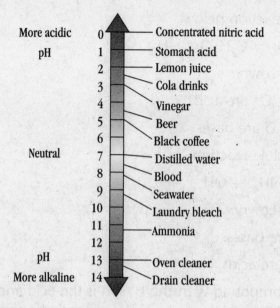

pH is actually just another way of expressing the concentration of H⁺ ions, [H⁺], in a solution. You can go back and forth between pH and [H⁺] by using the formulas below.

$$pH = -\log [H^+]$$

$$[H^+] = 10^{-pH}$$

For instance, if a solution has a hydrogen ion concentration, [H⁺], of 1×10^{-4} M, then the pH is 4. If a solution has [H⁺] equal to 1×10^{-11} M, then the pH is 11.

The reverse process works the same way. If a solution has a pH of 3, then the hydrogen ion concentration is 1×10^{-3} M. If a solution has a pH of 9, then [H⁺] is equal to 1×10^{-9} M.

It is important to notice that a *high* pH means a *low* [H⁺], which means that there are fewer H⁺ ions floating around and the solution is *less acidic*. Alternatively, a *low* pH means a *high* [H⁺], which means that there are more H⁺ ions floating around and the solution is *more acidic*.

Sometimes, it's more convenient to express acidity in terms of the hydroxide ion concentration, [OH⁻], by using pOH. You can use the same kinds of formulas for hydroxide ions as you can for hydrogen ions.

$$pOH = -\log [OH^-]$$

$$[OH^-] = 10^{-pOH}$$

You can think of acids and bases as being opposites. The greater the concentration of H⁺ ions in an aqueous solution, the smaller the concentration of OH⁻ ions, and vice versa. The formula below gives you this relationship.

$$pH + pOH = 14$$

That means that if you know the pOH of a solution, you can always find the pH, and vice versa. If the pOH of a solution is 5, then the pH must be 9. If the pH of a solution is 1, then the pOH must be 13.

WEAK ACIDS

Some acids do not fully dissociate when they are in water. That is, many of the H^+ ions don't dissociate from their conjugate base and the acid remains unionized. These are called **weak acids**. Because a weak acid produces a relatively small number of H^+ ions, it won't lower the pH as much as a strong acid like HCl will. Most of the acid molecules will remain in solution as undissociated aqueous particles. The same is true for **weak bases**, which don't dissociate very much and produce a small number of OH^- ions.

The dissociation constants, K_a and K_b, are measures of the strength of weak acids and bases. K_a and K_b, are just the equilibrium constants specific to acids and bases. You can make the equilibrium expressions for acids and bases the same way that you would make any other equilibrium expression, by making a fraction with the products in the numerator and the reactants and the denominator.

Acid dissociation constant

$$K_a = \frac{[H^+][B^-]}{[HB]}$$

$[H^+]$ = molar concentration of hydrogen ions (M)

$[B^-]$ = molar concentration of conjugate base ions (M)

$[HB]$ = molar concentration of undissociated acid molecules (M)

Base dissociation constant

$$K_b = \frac{[HB^+][OH^-]}{[B]}$$

$[HB^+]$ = conjugate acid ions (M)

$[OH^-]$ = molar concentration of hydroxide ions (M)

$[B]$ = base molecules (M)

Because an equilibrium expression always has the products in the numerator, you should be able to see that the greater the value of K_a, the greater the extent of the dissociation of the acid, the greater the amount of H^+ that is produced, and the stronger the acid. The same thing goes for K_b.

STRONG ACIDS

Strong acids produce a lot of H⁺ ions in water because they fully dissociate into H⁺ and a conjugate base anion. The same is true for strong bases, which completely dissociate to produce a lot of OH⁻ ions in water. You don't need to worry about the dissociation constants for strong acids and bases as long as you know they fully dissociate in water.

Important Strong Acids

HCl, HBr, HI, HNO_3, $HClO_4$, H_2SO_4

Important Strong Bases

$LiOH$, $NaOH$, KOH, $Ba(OH)_2$, $Sr(OH)_2$

BUFFERS

A **buffer** is a solution with a very stable pH. You can add acid or base to a buffer solution without greatly affecting the pH of the solution.

A buffer is created by placing a large amount of a weak acid or base into a solution along with its conjugate. A weak acid and its conjugate base can remain in solution together without neutralizing each other.

When both the acid and the conjugate base are together in the solution, any hydrogen ions that are added will be neutralized by the base while any hydroxide ions that are added will be neutralized by the acid without having much of an effect on the solution's pH.

POLYPROTIC ACIDS AND AMPHOTERIC SUBSTANCES

Some acids, such as H_2SO_4 and H_3PO_4, can give up more than one hydrogen ion. These are called **polyprotic** acids.

Substances that can act as either acids or bases are called **amphoteric** substances.

For instance, $H_2PO_4^-$ can act as an acid, giving up an H⁺ to become HPO_4^{2-}, or it can act as a base, accepting an H⁺ to become H_3PO_4.

HSO_4^- can act as an acid, giving up an H⁺ to become SO_4^{2-}, or it can act as a base, accepting an H⁺ to become H_2SO_4.

H_2O can act as an acid, giving up an H⁺ to become OH^-, or it can act as a base, accepting an H⁺ to become H_3O^+.

TITRATION

We learned earlier about neutralization reactions that occur between acids and bases to form water and a salt. A titration is simply a neutralization reaction. Titrations are used to determine the concentration of an acid or base solution. For example, if we had an unknown concentration of an acid solution, we would slowly add a known concentration and volume of a base solution until all the acid was used up. Because we know how much base was added, we can calculate how much acid must have been present. Neutralization reactions can be written in this form:

$$\text{Acid} + \text{Base} \rightarrow \text{Water} + \text{Salt}$$

The progress of a neutralization reaction can be shown in a titration curve. The diagram below shows the titration of a strong acid by a strong base.

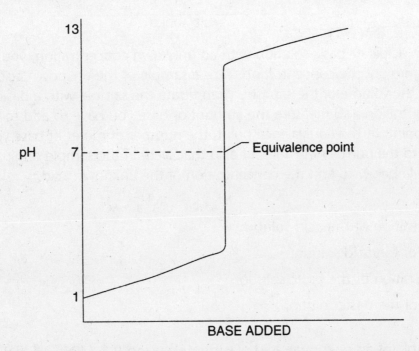

In this diagram, the pH increases slowly but steadily from the beginning of the titration until just before the equivalence point. The **equivalence point**, or **end point**, is the point in the titration when exactly enough base has been added to neutralize all the acid that was initially present. You use an **indicator** to tell when you have reached the equivalence point. Just before the equivalence point, the pH increases sharply as the last of the acid is neutralized. As more base is added beyond the equivalence point, the pH continues to increase as the solution becomes more basic.

The diagram on the next page shows the titration of a strong base with a strong acid. Notice that it looks like the previous diagram, except that it is reversed, with the pH decreasing instead of increasing.

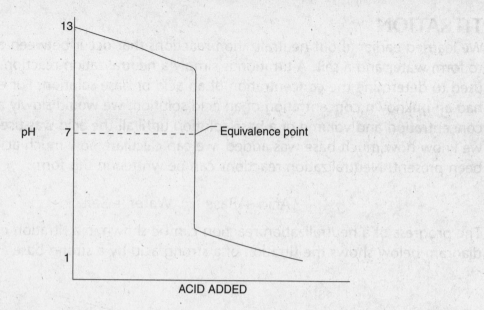

If you have an acidic or basic solution with an unknown concentration, you can use titration to figure out the concentration. Take a sample of the unknown acid solution, and measure the volume of the sample. Then titrate the sample with a basic solution of known concentration and measure the amount of base you have to add to reach the equivalence point. At the equivalence point, the number of moles of base you added will be equal to the number of moles of acid that were in the sample originally. You can use the formula below to find the concentration of the unknown acid.

$$M_A V_A = M_B V_B$$

M_A = concentration of the acid solution

V_A = volume of the acid solution

M_B = concentration of the basic solution

V_B = volume of the basic solution

A 50-mL sample of an unknown acid was titrated using 0.1 M NaOH. If 150 mL of the NaOH solution were required to reach the equivalence point, what was the concentration of the acid?

Use the formula.

$$M_A V_A = M_B V_B$$
$$(M_A)(50 \text{ mL}) = (0.1 \ M)(150 \text{ mL})$$
$$M_A = 0.3 \ M$$

So the concentration of the acid was 0.3 M.

REVIEW FOR LESSON 11

1. Which of the following statements is true of an acidic solution?
   A It has a pH greater than 7.
   B It has greater concentration of hydrogen ions than of hydroxide ions.
   C It has a slippery feel.
   D It will not conduct electricity.

2. Which of the following is true of a basic solution?
   F It has a pH equal to 7.
   G It has a sour taste.
   H It will conduct electricity.
   J It contains equal concentrations of hydrogen and hydroxide ions.

3. Which of the following compounds is the conjugate base of HNO_3?
   A H_2NO_3
   B HNO_3
   C NO_3^-
   D NH_3

4. If a solution has a pOH of 10, what is its pH?
   F 14
   G 10
   H 7
   J 4

5. If a solution has a hydrogen ion concentration of 1×10^{-7} M, what is the pH of the solution?
   A 1
   B 7
   C 10
   D 14

6. If a solution has a hydrogen ion concentration of 1.0×10^{-9} M, what is the pOH of the solution?
   F 1
   G 3
   H 5
   J 7

7. If a solution has a hydroxide ion concentration of 1.0×10^{-10} M, what is its pH?
   A 1
   B 2
   C 3
   D 4

Review for Lesson 11

Questions 8–12 are based on the table below.

Acid	Formula	Acid Dissociation Constant, K_a
Hydrofluoric acid	HF	6.8×10^{-4}
Hypochlorous acid	HClO	3.0×10^{-8}
Hydrocyanic acid	HCN	4.9×10^{-10}
Acetic acid	$HC_2H_3O_2$	1.8×10^{-5}

8 Which of the acids listed above is the strongest acid?

F Hydrofluoric acid
G Hypochlorous acid
H Hydrocyanic acid
J Acetic acid

9 Which of the acids listed above is the weakest acid?

A Hydrofluoric acid
B Hypochlorous acid
C Hydrocyanic acid
D Acetic acid

10 Which of the acids listed above is the strongest electrolyte?

F Hydrofluoric acid
G Hypochlorous acid
H Hydrocyanic acid
J Acetic acid

11 Which of the acids listed above is the weakest electrolyte?

A Hydrofluoric acid
B Hypochlorous acid
C Hydrocyanic acid
D Acetic acid

12 What is the expression for the dissociation constant, K_a, of hydrofluoric acid?

F $\dfrac{[HF]}{[H^+][F^-]}$

G $\dfrac{[H^+][F^-]}{[HF]}$

H $[H^+][F^-]$

J $[H^+][F^-][HF]$

Questions 13–16 are based on a titration experiment which produced the titration curve shown below.

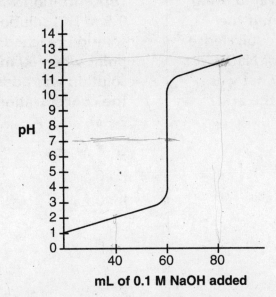

13. Which of the following statements is true about NaOH?
    A It is a strong acid.
    B It is a strong base.
    C It is a weak acid.
    D It is a weak base.

14. What is the pH of the solution when exactly 40 mL of NaOH solution has been added?
    F 2
    G 4
    H 7
    J 12

15. Approximately how many milliliters of NaOH solution must be added to bring the solution to the equivalence point?
    A 20 mL
    B 40 mL
    C 60 mL
    D 80 mL

16. Which of the following statements is true at the moment when 80 mL of NaOH solution has been added?
    F The pH is 12 and the solution is acidic.
    G The pH is 12 and the solution is basic.
    H The pH is 7 and the solution is acidic.
    J The pH is 7 and the solution is basic.

Review for Lesson 11

17 A 40.0-mL sample of an acid solution of unknown concentration was titrated using 0.20 M NaOH solution. If the solution reached the equivalence point when 50.0 mL of NaOH solution was added, what was the concentration of the acid?

   A 0.05 M
   B 0.10 M
   C 0.20 M
   D 0.25 M

18 A 30-mL sample of a basic solution of unknown concentration was titrated using 0.5 M HCl solution. If the solution reached the equivalence point when 60 mL of HCl solution was added, what was the concentration of the base?

   F 2 M
   G 1 M
   H 0.5 M
   J 0.2 M

ANSWERS AND EXPLANATIONS

1. **B is correct.** One of the definitions of an acid is that it is a substance that increases the concentration of H⁺ ions when you put it in water. An acidic solution always has a greater concentration of H⁺ ions than of OH⁻ ions. The other answers are wrong because an acidic solution will have a pH less than 7, it will conduct electricity, and it will not be slippery. Basic solutions are slippery. Did you use POE to cross off wrong answers?

2. **H is correct.** A basic solution contains hydroxide ions. Any solution that contains ions will conduct electricity. The other choices are wrong because a basic solution will have higher concentrations of OH⁻ ions than H⁺ ions and a pH of greater than 7. Basic solutions are slippery; acidic solutions are sour. Did you use POE to cross off wrong answers?

3. **C is correct.** The conjugate base of an acid is the part that remains when the H⁺ ion has been removed. After the H⁺ ion has been removed from HNO_3, NO_3^- remains, so NO_3^- is the conjugate base of HNO_3.

4. **J is correct.** Use the formula.

$$pH + pOH = 14$$

   If pOH is 10, then pH must be 4.

5. **B is correct.** Use the formula. You can use a calculator for this one, but you don't have to.

$$pH = -\log [H^+]$$

$$pH = -\log (1 \times 10^{-7}) = 7$$

6. **H is correct.** You need two steps here. First use the formula to find the pH.

$$pH = -\log [H^+]$$

$$pH = -\log (1.0 \times 10^{-9}) = 9$$

   Now remember the relationship between pH and pOH.

$$pH + pOH \times 14$$

   If pH is 9, pOH must be 5.

Review for Lesson 11

7   **D is correct.** You need two steps here. First, use the formula to find the pOH.

$$pOH = -\log [OH^-]$$

$$pOH = -\log (1.0 \times 10^{-10}) = 10$$

Now, remember the relationship between pH and pOH.

$$pH + pOH = 14$$

If pOH is 10, pH must be 4.

8   **F is correct.** The strongest acid is the one that dissociates the most to create the most H⁺ ions in solution. This acid will have the largest value of K_a. HF has the largest K_a because its exponent is the least negative. When in doubt on a question like this, guess either the largest or the smallest number listed.

9   **C is correct.** The weakest acid is the one that dissociates the least to create the fewest H⁺ ions in solution. It will have the smallest value of K_a. HCN has the smallest K_a because its exponent is the most negative. When in doubt on a question like this, guess either the largest or the smallest number listed.

10 **F is correct.** The strongest electrolyte is the substance that creates the most ions in a solution, so the strongest acid will be the strongest electrolyte. The strongest acid has the largest value of K_a. HF has the largest K_a because its exponent is the least negative. When in doubt on a question like this, guess either the largest or the smallest number listed.

11 **C is correct.** The weakest electrolyte is the substance that creates the least ions in a solution, so the weakest acid will be the weakest electrolyte. The weakest acid is the one that has the smallest value of K_a. HCN has the smallest K_a because its exponent is the most negative. When in doubt on a question like this, guess either the largest or the smallest number listed.

12 **G is correct.** The acid dissociation expression is an equilibrium expression. It is created by making a fraction with the concentrations of the products in the numerator, and the concentrations of the reactants in the denominator.

$$K_a = \frac{[H^+][F^-]}{[HF]}$$

13 **B is correct.** NaOH is on the list of strong bases that you should be familiar with. When NaOH is placed in a solution, it dissociates completely to produce OH⁻ ions.

14 F is correct. If you draw a line straight up from the 40-mL mark on the horizontal axis, it will cross the titration curve at the point where the pH is 2.

15 C is correct. The equivalence point occurs in the middle of the steep part of the titration curve. That's the point where enough base has been added to neutralize all of the acid that was originally present. When 60 mL of NaOH has been added on the curve in the question, the pH is exactly 7.

16 G is correct. If you draw a line straight up from the 80-mL mark on the horizontal axis, it will cross the titration curve at the point where the pH is 12. When the pH is greater than 7, the solution is basic. If you know that the pH of an acid must be less than 7, you can use POE to eliminate choices **F, H,** and **J**. This gets you the correct answer without even looking at the graph.

17 D is correct. Use the formula.
$$M_A V_A = M_B V_B$$
$$(M_A)(40.0 \text{ mL}) = (0.20 \text{ } M)(50.0 \text{ mL})$$
$$M_A = 0.25 \text{ } M$$
So the concentration of the acid was 0.25 M.

18 G is correct. Use the formula.
$$M_A V_A = M_B V_B$$
$$(0.5 \text{ } M)(60 \text{ mL}) = (M_B)(30 \text{ mL})$$
$$M_B = 1 \text{ } M$$
So the concentration of the base was 1 M.

LESSON 12
CAUSES OF CHEMICAL REACTIONS

COLLISION THEORY

Most reactions have two or more molecules coming together to form something new. To simplify things, you can think of a chemical reaction as occurring in two steps. In the first step, the bonds in the reactant molecules have to be broken. In the second step, the bonds of the product molecules are formed. We know that it takes energy to break a chemical bond and that energy is released when a bond is formed, so the first step is energetically uphill and the second step is downhill.

Collision theory is used to describe how chemical bonds are broken and made in chemical reactions. According to collision theory, reactants are constantly moving around and colliding with each other like pinballs. Each reactant molecule has a slightly different energy, some with more and some with less. Most collisions occur between reactant molecules that don't collide with enough energy to break bonds, but when two very energetic molecules collide, their bonds will break and the first step of a chemical reaction will begin. The energy required to break the reactant bonds is called the **activation energy, E_a**.

REACTION RATE

Imagine a reaction flask at the molecular level: Molecules are bouncing around constantly, hitting the walls of the flask and each other. When does the reaction actually take place? When two reactant molecules hit each other with enough energy to break a bond, the first step of a reaction will occur. Eventually, all the reactant molecules will collide with enough energy to break bonds and go on to become product molecules, and all the reactants will have been consumed. The rate at which reactants are consumed determines the rate of the reaction. Think about what affects the rate: It really comes down to how often the molecules collide, and what percentage of colliding molecules have enough energy to react. What factors do you think would change the reaction rate? If you increase the number of collisions going on, then their reaction rate will increase. If you increase the energy of the molecules, the reaction rate will also increase. Take a closer look:

Reaction rate increases with increasing concentration of reactants.

Higher concentrations of reactants means that there are more reactant molecules bouncing around in the space where the reaction takes place. More reactant molecules means more collisions. If there are more collisions, then there will be more reactions taking place. The more the reactants react, the faster the reaction goes.

Reaction rate increases with increasing temperature.

Increasing temperature means that the reactant molecules are moving faster, which means that the molecules have greater average kinetic energy. The higher the temperature, the greater the number of reactant molecules colliding with each other with enough energy (E_a) to cause a reaction.

A catalyst will increase the rate of a chemical reaction.

A catalyst increases the rate of a chemical reaction by providing a different way for reactant bonds to break that requires less energy. A catalyst speeds up a reaction by lowering the activation energy. A catalyst is not consumed in a reaction, and it doesn't appear in the balanced equation. Note that a catalyst *does not* change the equilibrium concentrations of reactants and products, it simply brings the reaction to equilibrium faster.

ENTHALPY

Enthalpy is a measure of the energy that is released or absorbed when bonds are broken and formed during a reaction.

When bonds are formed, energy is released.

Energy must be put into a bond in order to break it.

Remember what happens to bonds in a chemical reaction. First, energy must be put into the reactants to break their bonds. Once the reactant bonds are broken, product bonds can form. As the product bonds form, energy is released. If the products of a chemical reaction have stronger bonds than the reactants, then more energy is released in the making of product bonds than was put in to break the reactant bonds. In this case, energy is released overall and the reaction is **exothermic**. Another way of saying this is that the products are in a lower energy state than the reactants. Yet another way of saying this is that the change in enthalpy, ΔH, is negative.

If the products of a chemical reaction have weaker bonds than the reactants, then more energy is put in during the breaking of reactant bonds than is released in the making of product bonds. In this case, energy is absorbed overall and the reaction is **endothermic**. Another way of saying this is that the products are in a higher energy state than the reactants—the change in enthalpy, ΔH, is positive.

All substances like to be in the lowest possible energy state, which gives them the greatest stability. This means that, in general, exothermic processes are more likely to occur spontaneously than endothermic processes.

ENTROPY

Nature has a tendency to become increasingly disorganized. Think about spilling milk from a glass: Does the milk ever collect itself together and refill the glass? No, it spreads out randomly over the table and floor; in fact, it needed the glass in the first place just to have any shape at all. Likewise, think about helium in a balloon: it expands to fill its container, and if you empty the balloon, the helium diffuses randomly to fill the room. The reverse never happens: Helium molecules don't collect themselves from the atmosphere into a closed container. Just by looking around, you can see that it is the natural tendency of *all* things to increase their disorder. Disorder, or randomness, is measured as **entropy**. The greater the disorder of a system, the greater its entropy, ΔS.

While the overall inclination is to increase entropy, reactions can occur in which entropy decreases, but you must either put in energy or gain energy from making stable bonds. If randomness increases during the reaction, ΔS is positive for the reaction. For example, look at a decomposition reaction: H_2CO_3 breaks into H_2O and CO_2. In this case, one molecule breaks into two molecules and disorder is increased. The atoms are more organized in the carbonate molecule than they are as water and carbon dioxide molecules. If randomness decreases during the reaction, ΔS is negative. Look at the reverse reaction: If CO_2 and H_2O come together to form H_2CO_3, decreases entropy because the atoms in two molecules have become more organized by forming one molecule. Here are some other general examples.

- Liquids have more entropy than solids.
- Gases have more entropy than liquids.
- Particles in solution have more entropy than solids.
- Two moles of a substance have more entropy than one mole.

Because all substances like to increase their entropy, a reaction in which entropy is increasing is more likely to be spontaneous than a reaction in which entropy is decreasing. However, sometimes a substance's desire to increase its entropy as well as decrease its enthalpy results in a conflict. How is this resolved? When this happens, temperature acts as the referee. When temperature is low, decreasing enthalpy is more important. When temperature is high, increasing entropy is more important.

ENERGY DIAGRAMS

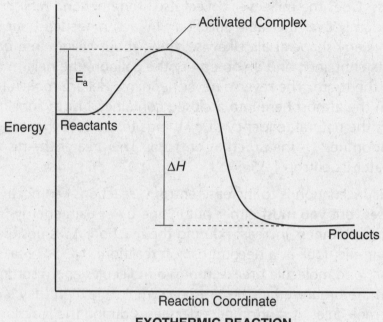

EXOTHERMIC REACTION

The diagram above shows the energy change that takes place during an exothermic reaction.

The reaction proceeds from left to right. On average, the reactants (on the left) start with a certain amount of energy. In order for the reaction to proceed, the reactants must have enough energy to reach the activated complex. This is the highest point on the graph above. The amount of energy needed to reach this point is called the activation energy, E_a. At this point, all reactant bonds have been broken, but no product bonds have been formed, so this is the point in the reaction with the highest energy and lowest stability.

Moving to the right, past the activated complex, product bonds start to form and eventually the energy level of the products is reached. This diagram represents an exothermic reaction, so the products are at a lower energy level than the reactants and ΔH is negative.

The following diagram is for an endothermic reaction.

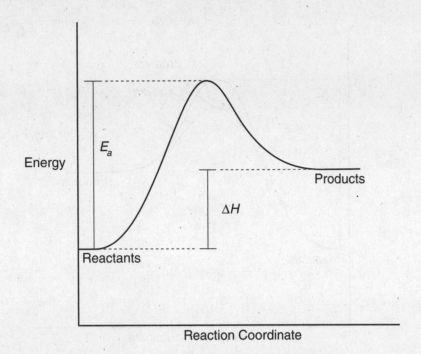

ENDOTHERMIC REACTION

This diagram differs from the diagram for the exothermic reaction because the energy of the products is greater than the energy of the reactants, so ΔH is positive.

Reaction diagrams can be read in both directions, so the reverse reaction for an exothermic reaction is endothermic and vice versa.

CATALYSTS AND ENERGY DIAGRAMS

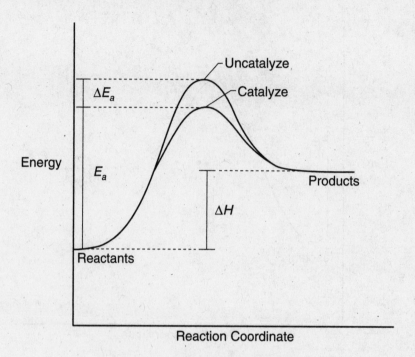

A catalyst speeds up a reaction by providing the reactants with an alternate pathway that has a lower activation energy, as shown in the diagram above.

Notice that the only difference between the catalyzed reaction and the uncatalyzed reaction is that the energy of the activated complex is lower for the catalyzed reaction. A catalyst lowers the activation energy, but it has no effect on the energy of the reactants, the energy of the products, or ΔH for the reaction. That means the equilibrium will be the same with or without the catalyst. That is, because the catalyst lowers the activation energy for both the forward and the reverse reaction, it has no effect on the equilibrium concentrations.

REVIEW FOR LESSON 12

1. In a chemical reaction, the activation energy is the energy—
   A required to break the bonds of the reactants
   B released when product bonds are formed
   C difference between the products and the reactants
   D of a gas molecule at standard temperature and pressure

2. When a catalyst is added to a chemical reaction, which of the following occurs?
   F The reaction will produce more products at equilibrium.
   G The reaction will produce more reactants at equilibrium.
   H The reaction will proceed to equilibrium more quickly.
   J The reaction will show a greater change in enthalpy.

3. Which of the following is always true of an exothermic reaction?
   A ΔH is positive for the reaction.
   B ΔH is negative for the reaction.
   C ΔS is positive for the reaction.
   D ΔS is negative for the reaction.

4. Which of the following phase changes results in an increase in entropy for the substance?
   F gas to liquid
   G gas to solid
   H liquid to solid
   J solid to liquid

5. Which of the following reactions results in an increase in randomness?
   A $Ag^+(aq) + Cl^-(aq) \rightarrow AgCl(s)$
   B $2\ Fe(s) + O_2(g) \rightarrow 2\ FeO(s)$
   C $H_2O(g) \rightarrow H_2O(l)$
   D $NaCl \rightarrow Na^+(aq) + Cl^-(aq)$

6. Which of the following changes to a reaction will always cause an increase in the rate of the reaction?
   F An increase in the temperature and an increase in the concentration of the reactants.
   G An increase in the temperature and a decrease in the concentration of the reactants.
   H A decrease in the temperature and an increase in the concentration of the reactants.
   J A decrease in the temperature and a decrease in the concentration of the reactants.

Review for Lesson 12

7 Which of the following is true of a reaction where ΔH is negative and ΔS is positive?

   A It is always spontaneous.
   B It is never spontaneous.
   C It is spontaneous only at high temperatures.
   D It is spontaneous only at low temperatures.

Questions 8–12 are based on the energy diagram for a reversible reaction shown below.

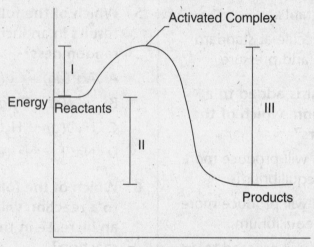

8 Which of the following represents the activation energy for the forward reaction?

   F I only
   G II only
   H I and II only
   J I, II, and III

9 Which of the following represents the activation energy for the reverse reaction?

   A I only
   B II only
   C III only
   D I and III only

10 The forward reaction represented in the diagram is—
   F an endothermic reaction
   G an exothermic reaction
   H both endothermic and exothermic
   J neither exothermic nor endothermic

11 Which of the following will be changed by the addition of a catalyst?
   A I only
   B II only
   C I and II only
   D I and III only

12 Which of the following represents the change in enthalpy for the forward reaction?
   F I only
   G II only
   H I and II only
   J II and III only

ANSWERS AND EXPLANATIONS

1. **A is correct.** The activation energy of a reaction is the energy required to break the bonds of the reactants and get the reaction started.

2. **H is correct.** A catalyst acts to speed up a chemical reaction, but it does not affect the equilibrium that the reaction eventually reaches. The catalyst speeds up the reaction by lowering the activation energy, but it does not affect the enthalpy change between reactants and products. Did you use POE?

3. **B is correct.** An exothermic reaction results in a decrease in enthalpy from the reactants to the products, so the change in enthalpy, ΔH, will be negative. You can't tell anything about the change in entropy, ΔS, from the fact that the reaction is exothermic. If you remembered that *exothermic* has something to do with ΔH, you could still use POE to eliminate choices **C** and **D**.

4. **J is correct.** Liquids are more random than solids, so a phase change from solid to liquid will produce an increase in entropy. Did you notice that three of the phase changes in the answer choices go in one direction and the correct answer goes in the other direction? If you are guessing, sometimes it's good to pick the answer that goes against the other three choices.

5. **D is correct.** In choice **A**, particles in solution combine to form a solid, thus decreasing randomness. In choice **B**, a solid combines with a gas to form a solid, and 3 moles come together to form 2 moles, so the randomness decreases. In choice **C**, a gas becomes a liquid, decreasing its randomness. In choice **D**, a solid breaks up into two particles in aqueous solution. Particles in solution have more randomness (greater entropy) than solid particles.

6. **F is correct.** An increase in temperature acts to increase the rate of a reaction and an increase in the concentration of reactants acts to increase the rate of a reaction. When these two are combined, the rate of a reaction will always increase. If you understand either the effect of temperature, or of concentration, but not both, you can still use POE to eliminate some wrong answer choices.

7. **A is correct.** If ΔH is negative, the reaction is exothermic. Exothermic reactions tend to be spontaneous. If ΔS is positive, randomness is increasing during the reaction, which will also tend to make the reaction spontaneous. Because both ΔH and ΔS favor spontaneity, the reaction will always be spontaneous.

8   **F is correct.** The activation energy of the reactants is represented by (I). That's the energy required to break the reactant bonds and start the reaction.

9   **C is correct.** The activation energy of the products is represented by (III). That's the energy required to break the product bonds so that the reaction can proceed in the reverse direction (right to left).

10 **G is correct.** The energy of the products is lower than the energy of the reactants, so energy must have been released during the reaction. When a reaction releases energy, it is exothermic.

11 **D is correct.** A catalyst lowers the activation energies of both the forward and reverse reactions, so both (I) and (III) will be changed.

12 **G is correct.** The difference between the energy of the products and the reactants is represented by (II). The energy of the products is less than the energy of the reactants, so the change in enthalpy, ΔH, is negative, and the reaction is exothermic.

LESSON 13
SCIENTIFIC INVESTIGATION

THE SCIENTIFIC METHOD

Experiments are central to all areas of science because they are the way that scientists ask and answer questions. Generally, experiments follow a basic method that starts with a question, suggests a theory or **hypothesis** about what the answer might be, and then uses an experiment to test that hypothesis. The **experimental design** must vary only *one* feature of the experiment at a time. That is, everything should be the same between the experiments except the **variable** being tested. Also, experiments are performed several times to ensure that the results are **repeatable**.

Here's an example. A scientist asks, "How will temperature affect the rate at which a particular reaction occurs?" The hypothesis might be that increasing the temperature will speed up the reaction. To test this, the scientist will perform an experiment where the reaction of interest occurs at several different temperatures and will record the rate of the reaction for each one. At the end, the scientist will examine the **data** to see how the temperature affected the rate of the reaction. Then, the experiment will be repeated in order to verify that the original experiment is repeatable.

Look at the experimental design. The scientist would be sure to run the reactions at the same concentrations of reactants, to stir them at the same rate, and to measure the rates in the same way. That is, the only variable would be the temperature, and all other factors would be held constant.

GRAPHS

Graphs are a visual way of looking at data. Data are often presented in visual form because it's easier to see patterns or trends by looking at a picture than by looking at a list of numbers. Some examples are given here.

LINEAR AND INCREASING

Typically, the data are presented in an "*x* versus *y*" format. In a linear increasing graph, the data increase by the same amount with each step. In this example, the increases are in equal steps. To help you see the usefulness of graphs, try plugging in numbers for *x* and *y*. For example, *x* could be the amount of money you have in your savings account, and *y* could be the number of weeks you've been saving. If you have been saving $5 every week, your total savings would have a linear relationship to time and would increase with time:

The data increases by the same amount with each step.

Data:

x	y
0	5
1	10
2	15
3	20
4	25

Graph:

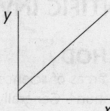

LINEAR AND DECREASING

Likewise, the data could have a downward trend and decrease. In this example, the data decrease by the same amount with each step. If you were taking money out of your savings every week to go to a movie, then the total balance would decrease. The balance would have a linear relationship to time and would decrease with time:

Data:

x	y
0	25
1	20
2	15
3	10
4	5

Graph:

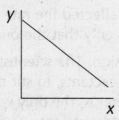

NONLINEAR AND INCREASING

Of course, the data are often more complicated and can increase in uneven steps. This is a nonlinear graph:

Data:

x	y
0	2
1	4
2	8
3	16
4	32

Graph:

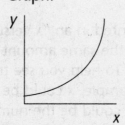

NONLINEAR AND DECREASING
A nonlinear decreasing trend could also occur. Here, the data decrease by different amounts with each step:

Data:

x	y
0	32
1	16
2	8
3	4
4	2

Graph:

SAFETY
Here's a rundown of some of the basic safety rules for the laboratory.

- A student should **never** work in a laboratory without qualified adult supervision.

- Don't put chemicals in your mouth.

- Always work with good ventilation; many common chemicals are toxic.

- When heating substances, do so slowly. When you heat things too quickly, they can spatter, burn, or explode.

- Don't weigh reactants directly on a scale. Use a glass or porcelain container to prevent corrosion of the balance pan.

ACCURACY
What good are experimental data if the experiment caused all sorts of errors? To be sure of accurate results, think about some things that might affect the experiment:

- When titrating, rinse the buret with the solution to be used in the titration instead of with water. If you rinse the buret with water, you might dilute the solution, which will cause the volume added from the buret to be too large.

- Don't contaminate your chemicals. Never insert another piece of equipment into a bottle containing a chemical. Instead, you should always pour the chemical into another clean container. Also, don't allow the inside of the stopper for a bottle containing a chemical touch another surface.

- When mixing chemicals, stir slowly to insure even distribution.

- Be conscious of significant figures when you record your results. The number of significant figures that you use should indicate the accuracy of your results.

- Allow hot objects to return to room temperature before weighing. Hot objects on a scale create convection currents, which may make the object seem lighter than it is.

- When collecting a gas over water, remember to take into account the pressure and volume of the water vapor.

REVIEW FOR LESSON 13

Questions 1–3 are based on the information given below.

A scientist conducted an experiment to evaluate the effects of temperature and reactant concentration on the speed of the following reaction.

$$A + B \rightarrow C$$

The data for the experiment are shown below.

Trial	Concentration of A	Concentration of B	Temperature	Reaction Rate
1	1 M	1 M	300 K	0.1 M/sec.
2	1 M	2 M	300 K	0.2 M/sec.
3	2 M	1 M	300 K	0.4 M/sec.
4	1 M	1 M	350 K	0.5 M/sec.
5	2 M	2 M	350 K	3.0 M/sec.

1 Which two trials can be used to examine the effect of changing the concentration of reactant A?

   A Trial 1 and Trial 2
   B Trial 1 and Trial 3
   C Trial 1 and Trial 4
   D Trial 1 and Trial 5

2 Which two trials can be used to examine the effect of changing the concentration of reactant B?

   F Trial 1 and Trial 2
   G Trial 1 and Trial 3
   H Trial 1 and Trial 4
   J Trial 1 and Trial 5

3 Which two trials can be used to examine the effect of changing the temperature?

   A Trial 1 and Trial 2
   B Trial 1 and Trial 3
   C Trial 1 and Trial 4
   D Trial 1 and Trial 5

4 The data below were recorded in an experiment.

Time (min.)	Pressure (mm Hg)
1	10
2	20
3	30
4	40

Which of the graphs below best shows the data?

F
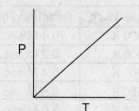

H
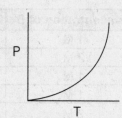

G
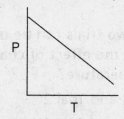

J
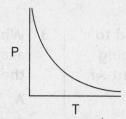

5 The data below were recorded in an experiment.

Time (min.)	Pressure (mm Hg)
1	100
2	200
3	400
4	800

Which of the graphs below best shows the data?

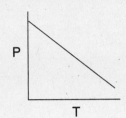
A

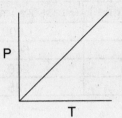
C

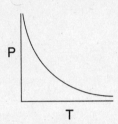
B

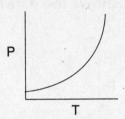
D

6 The data below were recorded in an experiment.

Time (min.)	Pressure (mm Hg)
1	200
2	150
3	100
4	50

Which of the graphs below best shows the data?

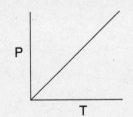
F

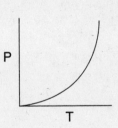
H

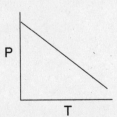
G

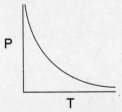
J

7 The data below were recorded in an experiment.

Time (min.)	Pressure (mm Hg)
1	1,000
2	500
3	250
4	125

Which of the graphs below best shows the data?

A
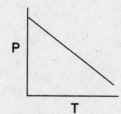

C
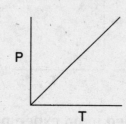

B
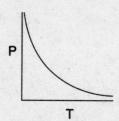

D
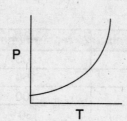

ANSWERS AND EXPLANATIONS

1. **B is correct.** To examine the effect of changing the concentration of A, you need to compare two trials in which everything stays constant except the concentration of A. That's 1 and 3.

2. **F is correct.** To examine the effect of changing the concentration of B, you need to compare two trials in which everything stays constant except the concentration of B. That's 1 and 2.

3. **C is correct.** To examine the effect of changing the temperature, you need to compare two trials in which everything stays constant except the temperature. That's 1 and 4.

4. **F is correct.** The numbers shown increase by the same amount with each interval, so the graph is an increasing straight line.

5. **D is correct.** The numbers shown increase by different amounts with each interval, so the graph is an increasing curved line.

6. **G is correct.** The numbers shown decrease by the same amount with each interval, so the graph is a decreasing straight line.

7. **H is correct.** The numbers shown decrease by different amounts with each interval, so the graph is a decreasing curved line.

THE PRACTICE TESTS

INSTRUCTIONS FOR TAKING THE PRACTICE TESTS

Congratulations! If you've made it this far, you've covered every Standard of Learning (SOL) that the Virginia EOC Chemistry exam will test you on. Now it's time to show what you know with the two practice EOC Chemistry exams. Now is the time to brush up on any SOL or question-type you may be rusty on.

Each practice test is 60 questions long, just like the real exam. The test is untimed, but you do have to take in one sitting. Take the entire test at once!

Here are the materials you should have before taking one of the practice tests in this book:

- a four-function calculator
- a ruler
- scrap paper
- the Periodic Table of Elements (found at the beginning of each practice test)

To record your answers, fill in the appropriate bubbles on the answer sheet for each test. You can find the answer sheets before each exam.

The correct answer and a detailed explanation for every question follow each exam. Use the explanations to help you prepare for the real exam.

Practice Test 1

Periodic Table of Elements

1 H 1.0																	2 He 4.0
3 Li 6.9	4 Be 9.0											5 B 10.8	6 C 12.0	7 N 14.0	8 O 16.0	9 F 19.0	10 Ne 20.2
11 Na 23.0	12 Mg 24.3											13 Al 27.0	14 Si 28.1	15 P 31.0	16 S 32.1	17 Cl 35.5	18 Ar 39.9
19 K 39.1	20 Ca 40.1	21 Sc 45.0	22 Ti 47.9	23 V 50.9	24 Cr 52.0	25 Mn 54.9	26 Fe 55.8	27 Co 58.9	28 Ni 58.7	29 Cu 63.5	30 Zn 65.4	31 Ga 69.7	32 Ge 72.6	33 As 74.9	34 Se 79.0	35 Br 79.9	36 Kr 83.8
37 Rb 85.5	38 Sr 87.6	39 Y 88.9	40 Zr 91.2	41 Nb 92.9	42 Mo 95.9	43 Tc (98)	44 Ru 101.	45 Rh 102.	46 Pd 106.4	47 Ag 107.	48 Cd 112.	49 In 114.8	50 Sn 118.	51 Sb 121.	52 Te 127.6	53 I 126.	54 Xe 131.
55 Cs 132.	56 Ba 137.3	57 *La 138.	72 Hf 178.5	73 Ta 180.	74 W 183.9	75 Re 186.	76 Os 190.2	77 Ir 192.	78 Pt 195.1	79 Au 197.	80 Hg 200.6	81 Tl 204.4	82 Pb 207.	83 Bi 209.0	84 Po (209)	85 At (210)	86 Rn (222)
87 Fr (223)	88 Ra 226.0	89 †Ac 227.	104 Unq (261)	105 Unp (262)	106 Unh (263)	107 Uns (262)	108 Uno (265)	109 Une (267)									

*Lanthanide Series

58 Ce 140.	59 Pr 140.9	60 Nd 144.	61 Pm (145)	62 Sm 150.	63 Eu 152.	64 Gd 157.3	65 Tb 158.	66 Dy 162.	67 Ho 164.9	68 Er 167.	69 Tm 168.	70 Yb 173.0	71 Lu 175.

†Actinide Series

90 Th 232.0	91 Pa (231)	92 U 238.0	93 Np (237)	94 Pu (244)	95 Am (243)	96 Cm (247)	97 Bk (247)	98 Cf (251)	99 Es (252)	100 Fm (257)	101 Md (258)	102 No (259)	103 Lr (260)

PRACTICE TEST 1 ANSWER SHEET

Name: _____

1. Ⓐ Ⓑ Ⓒ Ⓓ
2. Ⓕ Ⓖ Ⓗ Ⓙ
3. Ⓐ Ⓑ Ⓒ Ⓓ
4. Ⓕ Ⓖ Ⓗ Ⓙ
5. Ⓐ Ⓑ Ⓒ Ⓓ
6. Ⓕ Ⓖ Ⓗ Ⓙ
7. Ⓐ Ⓑ Ⓒ Ⓓ
8. Ⓕ Ⓖ Ⓗ Ⓙ
9. Ⓐ Ⓑ Ⓒ Ⓓ
10. Ⓕ Ⓖ Ⓗ Ⓙ
11. Ⓐ Ⓑ Ⓒ Ⓓ
12. Ⓕ Ⓖ Ⓗ Ⓙ
13. Ⓐ Ⓑ Ⓒ Ⓓ
14. Ⓕ Ⓖ Ⓗ Ⓙ
15. Ⓐ Ⓑ Ⓒ Ⓓ
16. Ⓕ Ⓖ Ⓗ Ⓙ
17. Ⓐ Ⓑ Ⓒ Ⓓ
18. Ⓕ Ⓖ Ⓗ Ⓙ
19. Ⓐ Ⓑ Ⓒ Ⓓ
20. Ⓕ Ⓖ Ⓗ Ⓙ
21. Ⓐ Ⓑ Ⓒ Ⓓ
22. Ⓕ Ⓖ Ⓗ Ⓙ
23. Ⓐ Ⓑ Ⓒ Ⓓ
24. Ⓕ Ⓖ Ⓗ Ⓙ
25. Ⓐ Ⓑ Ⓒ Ⓓ
26. Ⓕ Ⓖ Ⓗ Ⓙ
27. Ⓐ Ⓑ Ⓒ Ⓓ
28. Ⓕ Ⓖ Ⓗ Ⓙ
29. Ⓐ Ⓑ Ⓒ Ⓓ
30. Ⓕ Ⓖ Ⓗ Ⓙ
31. Ⓐ Ⓑ Ⓒ Ⓓ
32. Ⓕ Ⓖ Ⓗ Ⓙ
33. Ⓐ Ⓑ Ⓒ Ⓓ
34. Ⓕ Ⓖ Ⓗ Ⓙ
35. Ⓐ Ⓑ Ⓒ Ⓓ
36. Ⓕ Ⓖ Ⓗ Ⓙ
37. Ⓐ Ⓑ Ⓒ Ⓓ
38. Ⓕ Ⓖ Ⓗ Ⓙ
39. Ⓐ Ⓑ Ⓒ Ⓓ
40. Ⓕ Ⓖ Ⓗ Ⓙ
41. Ⓐ Ⓑ Ⓒ Ⓓ
42. Ⓕ Ⓖ Ⓗ Ⓙ
43. Ⓐ Ⓑ Ⓒ Ⓓ
44. Ⓕ Ⓖ Ⓗ Ⓙ
45. Ⓐ Ⓑ Ⓒ Ⓓ
46. Ⓕ Ⓖ Ⓗ Ⓙ
47. Ⓐ Ⓑ Ⓒ Ⓓ
48. Ⓕ Ⓖ Ⓗ Ⓙ
49. Ⓐ Ⓑ Ⓒ Ⓓ
50. Ⓕ Ⓖ Ⓗ Ⓙ
51. Ⓐ Ⓑ Ⓒ Ⓓ
52. Ⓕ Ⓖ Ⓗ Ⓙ
53. Ⓐ Ⓑ Ⓒ Ⓓ
54. Ⓕ Ⓖ Ⓗ Ⓙ
55. Ⓐ Ⓑ Ⓒ Ⓓ
56. Ⓕ Ⓖ Ⓗ Ⓙ
57. Ⓐ Ⓑ Ⓒ Ⓓ
58. Ⓕ Ⓖ Ⓗ Ⓙ
59. Ⓐ Ⓑ Ⓒ Ⓓ
60. Ⓕ Ⓖ Ⓗ Ⓙ

1. An atom of chlorine-37 contains—
   A 17 protons and 17 neutrons
   B 17 protons and 20 neutrons
   C 20 protons and 17 neutrons
   D 37 protons and 37 neutrons

2. Which of the following is the electron configuration of a nitrogen atom?
   F $1s^2\ 2s^2$
   G $1s^2\ 2s^2 2p^3$
   H $1s^2\ 2s^2 2p^5$
   J $1s^2\ 2s^2 2p^6\ 3s^2$

3. Which of the following statements is true of an electron?
   A An electron is positively charged and smaller than a proton.
   B An electron is negatively charged and larger than a proton.
   C An electron is positively charged and larger than a proton.
   D An electron is negatively charged and smaller than a proton.

4. A scientist examined four substances in a laboratory. Which of the following is most likely a metal?
   F A liquid that is a poor conductor of electricity.
   G A shiny solid with a low ionization energy.
   H A crystalline solid with a high ionization energy.
   J A gas with a high electronegativity.

5. Which of the following elements has the highest electronegativity?
   A B
   B C
   C N
   D O

6. Which of the following compounds contains a covalent bond?
   F Mg_3N_2
   G Li_2O
   H CO_2
   J $CaBr_2$

Practice Test 1

7 The oxidation state of phosphorous is greatest in which of the following compounds?

A P_2O_4

B PCl_5

C PCl_3

D Li_3P

8 What is the name of the compound represented by NBr_3?

F Nitrogen bromide

G Bromine nitrogen

H Nitrogen tribromine

J Nitrogen tribromide

9 What is the formula of iron(III) chloride?

A FeCl

B Fe_3Cl

C $FeCl_3$

D Fe_3Cl_3

10 What is the oxidation state of oxygen in an O_2 molecule?

F 0

G +1

H +2

J −2

11 Which of the following is a physical change?

A Ice melting

B Paper burning

C Milk going sour

D Meat cooking

12 ___Na + ___H_2O → ___NaOH + ___H_2

When the equation above is properly balanced, what is the coefficient for NaOH?

F 8

G 6

H 4

J 2

13 Which of the following is a combustion reaction?

A $2 C_4H_6 + 7 O_2 \rightarrow 6 H_2O + 8 CO_2$

B $HNO_3 + NaOH \rightarrow H_2O + NaNO_3$

C $Mn + Ag_2SO_4 \rightarrow MnSO_4 + 2 Ag$

D $2 NiO \rightarrow 2 Ni + O_2$

14 Which of the following shows the conversion of 1230 grams to kilograms?

F $(1{,}230 \text{ g})\left(\dfrac{1{,}000 \text{ g}}{1 \text{ kg}}\right) =$

G $(1{,}230 \text{ g})\left(\dfrac{1 \text{ kg}}{1{,}000 \text{ g}}\right) =$

H $(1{,}230 \text{ g})\left(\dfrac{10 \text{ g}}{1 \text{ kg}}\right) =$

J $(1{,}230 \text{ g})\left(\dfrac{10 \text{ kg}}{1 \text{ g}}\right) =$

15 A student converted 200 centimeters into meters. Which of the following shows the student's answer with the correct number of significant digits?

A 2.000 meters

B 2.00 meters

C 2.0 meters

D 2 meters

16 One mole of a compound was found to have a mass of 32 grams. Which of the following could be the identity of the compound?

F O_2

G N_2

H H_2

J Cl_2

17 A balloon contains 5.6 liters of helium at standard temperature and pressure. How many helium atoms does the balloon contain?

A 7.5×10^{23} helium atoms

B 6.0×10^{23} helium atoms

C 4.0×10^{23} helium atoms

D 1.5×10^{23} helium atoms

18 Which of the following statements is true about a 0.100 mole sample of SO_2 gas at STP?

F The sample occupies 22.4 liters and weighs 64.1 grams.

G The sample occupies 2.24 liters and weighs 64.1 grams.

H The sample occupies 22.4 liters and weighs 6.41 grams.

J The sample occupies 2.24 liters and weighs 6.41 grams.

19 The following reaction took place at STP:

$$C_2H_2(g) + 5\,O_2(g) \rightarrow 4\,CO_2(g) + 2\,H_2O(g)$$

How many liters of CO_2 gas were produced if 2.50 moles of O_2 were consumed in the reaction?

A 11.2 L

B 22.4 L

C 44.8 L

D 89.6 L

Go On

20 In a chemical reaction, the limiting reactant is the reactant that—

F is present in the smallest molar quantity

G occupies the smallest volume

H disappears first in the reaction

J has the smallest molecular weight

21 How many moles of oxygen atoms are present in 1 mole of $Ca(NO_3)_2$?

A 1

B 2

C 3

D 6

22 A sample of nitrogen gas in a 10 liter container was cooled from 47.0°C to 27.0°C. If the initial pressure of the gas was 848 mm Hg, what was the final pressure?

F 905 mm Hg

G 795 mm Hg

H 487 mm Hg

J 244 mm Hg

23 Absolute zero is the temperature at which all motion stops. Which of the following temperatures is absolute zero?

A 100 K

B 0° C

C 0 K

D 273°C

24 Which of the following is true about two gas samples that occupy equal volumes at the same temperature and pressure?

F The two samples must both be H_2.

G The two samples must contain the same number of moles.

H The two samples must have the same weight.

J The temperature of both samples must be greater than 100°C.

25 Which of the following lists shows intermolecular forces in order of decreasing strength?

A Dipole-dipole interactions, London dispersion forces, hydrogen bonding

B Dipole-dipole interactions, hydrogen bonding, London dispersion forces

C Hydrogen bonding, London dispersion forces, dipole-dipole interactions

D Hydrogen bonding, dipole-dipole interactions, London dispersion forces

Go On

Questions 26 and 27 are based on the diagram below.

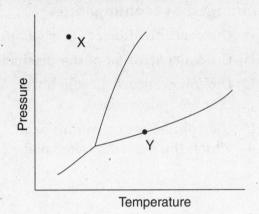

26 Which phase of matter is represented by Point X on the diagram above?
F Solid
G Liquid
H Gas
J Plasma

27 The phase change shown at Point Y could be—
A sublimation
B melting
C boiling
D freezing

28 The specific heat capacity of water is 4.2 J/g-°C. How much heat is required to raise the temperature of a 75 gram sample of water from 10°C to 22°C?
F 3780 J
G 3150 J
H 214 J
J 26 J

Questions 29 and 30 are based on the diagram below.

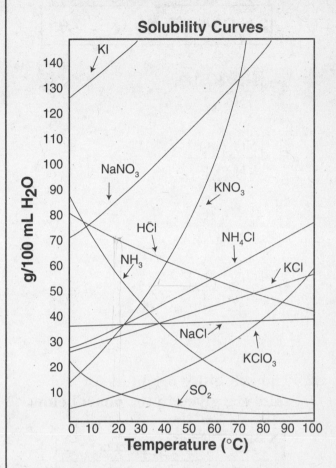

29 Approximately how many grams of HCl can be dissolved in 100 mL of water at a temperature of 20°C?
A 110 g
B 70 g
C 40 g
D 20 g

30 Which of the substances listed below is the most soluble at a temperature of 80°C?
F NH_3
G KCl
H NaCl
J NH_4Cl

Practice Test 1

31 Write the equilibrium expression for the reaction shown below.

$$2\ N_2O_5(g) \rightleftarrows 4\ NO_2(g) + O_2(g)$$

A $\dfrac{[NO_2][O_2]}{[N_2O_5]}$

B $\dfrac{[NO_2]^4[O_2]}{[N_2O_5]^2}$

C $\dfrac{4[NO_2][O_2]}{2[N_2O_5]}$

D $\dfrac{[NO_2]^2[O_2]}{[N_2O_5]}$

32 The solubility products for some salts are given in the chart below.

Salt	Solubility product, K_{sp}
$Cu(OH)_2$	2×10^{-20}
$Fe(OH)_2$	2×10^{-14}
$Mg(OH)_2$	1×10^{-11}
$Zn(OH)_2$	2×10^{-14}

Which of the salts listed in the chart is the most soluble?

F $Cu(OH)_2$
G $Fe(OH)_2$
H $Mg(OH)_2$
J $Zn(OH)_2$

33 The value of the equilibrium constant for a reaction can be changed by a change in—

A the concentration of the reactants
B the concentration of the products
C the temperature at which the reaction takes place
D the volume of the container in which the reaction takes place

34 The reversible reaction shown below is at equilibrium in a closed container.

$$SO_2(g) + NO_2(g) \rightarrow SO_3(g) + NO(g)$$

Which of the following changes to the equilibrium conditions will bring about an increase in the concentration of NO gas at equilibrium?

F The addition of SO_2 gas
G The removal of NO_2 gas
H The addition of SO_3 gas
J The removal of SO_2 gas

35 Which of the following statements is true of an acidic solution?

A It has a pH less than 7 and an H^+ concentration greater than $10^{-7}\ M$.
B It has a pH greater than 7 and an H^+ concentration greater than $10^{-7}\ M$.
C It has a pH less than 7 and an H^+ concentration less than $10^{-7}\ M$.
D It has a pH greater than 7 and an H^+ concentration less than $10^{-7}\ M$.

36 If a solution has a pH of 8, what is its pOH?

F 14
G 12
H 8
J 6

Questions 37 and 38 are based on a titration experiment that produced the titration curve shown below.

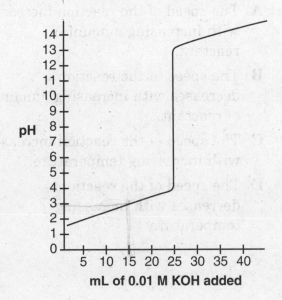

mL of 0.01 M KOH added

37 What is the pH of the solution when exactly 15 mL of KOH solution has been added?

A 2.5
B 4.0
C 5.5
D 7.0

38 Approximately how many milliliters of KOH solution must be added to bring the solution to the equivalence point?

F 50 mL
G 40 mL
H 35 mL
J 25 mL

39 Which of the following phase changes results in a decrease in randomness over the course of the change?

A Gas to liquid
B Liquid to gas
C Solid to gas
D Solid to liquid

40 Which of the following changes to a reaction will always cause a decrease in the rate of the reaction?

F An increase in the temperature and an increase in the concentration of the reactants.
G An increase in the temperature and a decrease in the concentration of the reactants.
H A decrease in the temperature and an increase in the concentration of the reactants.
J A decrease in the temperature and a decrease in the concentration of the reactants.

Go On

41 Which of the following is true of a reaction in which ΔH is positive and ΔS is negative?

A It is always spontaneous.

B It is never spontaneous.

C It is spontaneous only at high temperatures.

D It is spontaneous only at low temperatures.

42 In the reaction shown below, two hydrogen atoms combine to form an isotope of helium.

$$^{2}_{1}H + ^{1}_{1}H \rightarrow ^{3}_{2}He$$

This is an example of—

F combustion

G oxidation-reduction

H fusion

J fission

43 The chart below shows data collected for an experiment.

Trial	Amount of reaction	Temp.	Speed of the reactant
1	1 mole	300 K	0.2 moles per second
2	1 mole	310 K	0.3 moles per second
3	1 mole	320 K	0.5 moles per second
4	1 mole	330 K	0.8 moles per second

Which of the following statements is best supported by the experimental data?

A The speed of the reaction increases with increasing amount of reactant.

B The speed of the reaction decreases with increasing amount of reactant.

C The speed of the reaction increases with increasing temperature.

D The speed of the reaction decreases with increasing temperature.

Go On

44 The data below were recorded in an experiment.

Time (min.)	Pressure (mm Hg)
1	180
2	160
3	140
4	120

Which of the graphs below best shows the data?

F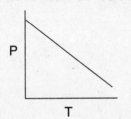

H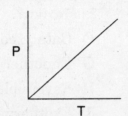

G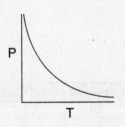

J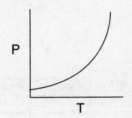

45 When a metal forms an ionic bond with a nonmetal, the metal atom will—

A lose an electron and become a negative ion

B lose an electron and become a positive ion

C gain an electron and become a negative ion

D gain an electron and become a positive ion

46 A student dissolved a block of salt in a beaker of water. The student found that the temperature of the water in the beaker increased in the process. Which of the following is true?

F The dissolution of the salt released energy and is endothermic.

G The dissolution of the salt released energy and is exothermic.

H The dissolution of the salt absorbed energy and is endothermic.

J The dissolution of the salt absorbed energy and is exothermic.

47 Which of the following ions has the *same* electron configuration as argon?

A K^+

B Ag^+

C Na^+

D Cu^+

48 When a piece of copper is placed in a beaker of water, the copper will sink to the bottom of the beaker. This is because—

F copper has a greater specific heat than water

G copper has a greater boiling point than water

H copper has a greater electronegativity than water

J copper has a greater density than water

49. Which of the following is true of the elements B, C, N, and O?

   A They have the same number of valence electrons.
   B They have the same number of protons.
   C They have the same number of electron shells.
   D They have the same molecular weights.

50. A catalyst increases the rate of a chemical reaction by—

   F decreasing the concentration of reactants
   G increasing the concentration of reactants
   H decreasing the activation energy for the reaction
   J increasing the activation energy for the reaction

51. Elements from group 14 and 15 would *most* likely form what type of bond?

   A Covalent
   B Ionic
   C Metallic
   D Hydrogen

52.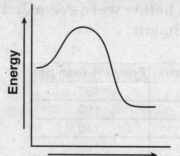

   The potential energy diagram above shows the progress of a reaction. What type of reaction is shown?

   F Catalyzed
   G Endothermic
   H Reversible
   J Exothermic

53.
Sample	Protons	Neutrons	Electrons
1	19	20	18
2	19	20	19
3	19	21	20
4	19	22	18

Which of these isotopes is the heaviest?

   A 1
   B 2
   C 3
   D 4

54. A specific mass of potassium bromide is needed for an experiment. Which of the following pieces of equipment would measure the mass with the greatest accuracy?

F Scale
G Triple beam balance
H Graduated cylinder
J Barometer

55. Which of the following diagrams indicates a decrease in entropy?

A
Ice melting

B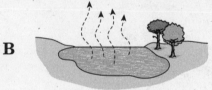
Water evaporating

C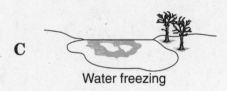
Water freezing

D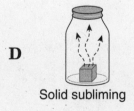
Solid subliming

56. $\_\_Al(s) + \_\_O_2(g) \rightarrow \_\_Al_2O_3(s)$

Aluminum oxide is formed from the reaction of solid aluminum and oxygen gas. If the reaction uses 4 moles of solid aluminum, how many moles of oxygen are needed?

F 1
G 2
H 3
J 4

57. $R = 0.0821 \ [L \cdot atm/K \cdot mol]$

The ideal gas constant is experimentally determined. A container is filled with 30.0 moles of carbon dioxide at 300.0 K. If the volume of the container is 1.00 L, what is the pressure of the gas?

A 746 atm
B 739 atm
C 821 atm
D 900 atm

58 A thermometer measures the temperature to the nearest 0.1°C. A student uses this thermometer to measure the temperature of a solution as 87.65°C. Which of the digits is the least accurate?

F 8
G 7
H 6
J 5

59 A student makes a solution using 15.6 g of sodium chloride in 300.0 mL of water. What is the molarity of the solution?

A 0.650 M
B 0.890 M
C 0.950 M
D 1.780 M

60 M^{+2}

Which element forms the ions shown above?

F Calcium
G Lithium
H Oxygen
J Potassium

Answers and Explanations for Practice Test 1

ANSWERS AND EXPLANATIONS FOR PRACTICE TEST 1

PRACTICE TEST 1 ANSWERS

1. **B is correct.** The number of protons in an atom is given by the atomic number (17 in this case). You can get the number of neutrons by subtracting the atomic number from the mass number (37 − 17 = 20).

2. **G is correct.** Nitrogen has the atomic number seven, so you know that nitrogen contains a total of 7 electrons. Only choice **G** has electron numbers that add up to 7 (2 + 2 + 3 = 7)

3. **D is correct.** Electrons are negatively charged, and they are much smaller than protons. By the way, protons are positively charged and neutrons have no charge. If you know either the size or the charge of an electron, but not both, you can still use POE to eliminate some wrong answers.

4. **G is correct.** Metals are typically found as shiny solids that are good conductors of heat and electricity. They have low ionization energies and low electronegativities. Did you use POE here?

5. **D is correct.** Electronegativity increases as you move from left to right across a period on the periodic table. O (oxygen) is the farthest to the right of the choices listed.

6. **H is correct.** Covalent bonds take place between nonmetals. CO_2 is the only choice that shows two nonmetals bonded together.

7. **B is correct.** Chlorine takes an oxidation state of −1, so in PCl_5, phosphorous takes a +5 oxidation state. In PCl_3 (choice **C**), phosphorous takes a +3 oxidation state. In P_2O_4 (choice **A**), oxygen takes a −2 oxidation state, so phosphorous must take an oxidation state of +4. In Li_3P (choice **D**), lithium takes a +1 oxidation state, so phosphorous must take a −3 oxidation state. So phosphorous takes the greatest oxidation state (+5) in PCl_5.

8. **J is correct.** Nitrogen and bromine are nonmetals, so NBr_3 is a molecular compound. You use prefixes to show the number of atoms of each element when you name a molecular compound. You also change the ending of the second element to -ide. So NBr_3 is nitrogen tribromide.

9. **C is correct.** In an ionic compound that contains a transition metal, a Roman number is used to show the oxidation state of the metal. If iron (Fe) has an oxidation state of +3 in the compound, then there must be three chlorine atoms, because chlorine always takes a −1 oxidation state.

10. **F is correct.** The oxidation state of an element that is not combined with another element is zero. So the oxidation state of oxygen in O_2 is zero.

11 A is correct. The melting of water is a phase change, because the molecular formula of water, H_2O, is not changed when water melts. All of the other changes listed take place because of chemical reactions, so they are chemical changes.

12 J is correct. Backsolve on this one. If there are 2 NaOH molecules, then there must be 2 Na atoms and 2 H_2O molecules. Now you can count the H atoms on both sides of the reaction. There are 4 H atoms on the left side and 2 H atoms on the right side, so if you put 2 H_2 molecules on the right, the equation will be balanced.

$$2\ Na + 2\ H_2O \rightarrow 2\ NaOH + H_2$$

If you chose any of the other answer choices, you found whole number coefficients but not the *lowest* whole number coefficients. You need lowest whole number coefficients to balance an equation properly.

13 A is correct. In choice **A**, a carbon compound reacts with oxygen to form water and carbon dioxide. That's combustion. Choice **B** is a neutralization reaction. Choice **C** is a single replacement reaction. Choice **D** is a decomposition reaction.

14 G is correct. In choice **G**, 1230 g is multiplied by a fraction equal to 1, with grams in the denominator so that the only units left at the end will be kg. By the way, 1,230 g is equal to 1.23 kg.

15 D is correct. The conversion from centimeters to meters is shown below.

$$(200\ cm)\left(\frac{1\ m}{100\ cm}\right) = \frac{200}{100} m = 2m$$

There is only one significant digit in the number 200, so the answer in meters can only have one significant digit.

16 F is correct. If one mole of a compound has a mass of 32 grams, then the molecular weight of the compound must be 32 g/mole. The periodic table tells us that the molecular weights of the four answer choices are: O_2—32, N_2—28, H_2—2, Cl_2—71.

17 D is correct. This is a two-step problem. To convert from liters to atoms, first convert from liters to moles and then from moles to atoms.

$$5.6 \text{ L} \times \left(\frac{\text{mole}}{22.4 \text{ L}}\right) = 0.25 \text{ mole}$$

$$0.25 \text{ mole} \times \left(\frac{6.02 \times 10^{23} \text{ atoms}}{1 \text{ mole}}\right) = 1.5 \times 10^{23} \text{ atoms}$$

18 J is correct. Find the volume of the gas by using the STP conversion formula.

$$\text{Moles} = \frac{\text{liters}}{(22.4 \text{ L}/\text{mole})}$$

Liters = (moles)(22.4 L/mole)

$$\text{Liters} = \left(\frac{0.100 \text{ mole}}{1}\right)\left(\frac{22.4 \text{ L}}{1 \text{ mole}}\right) = 2.24 \text{ liters of } SO_2$$

Next, find the molecular weight of SO_2 from the periodic table. Then, use the formula to convert from moles to grams.

32.1 g/mole + 2(16.0 g/mole) = 64.1 g/mole

$$\text{Moles} = \frac{\text{grams}}{\text{molecular weight}}$$

Grams = (moles)(molecular weight)

Grams = (0.100 moles)(64.1 g/mole) = 6.41 grams of SO_2.

If you know about either volume or mass, but not both, you can still use POE to eliminate some wrong answers.

Answers and Explanations for Practice Test 1

19 C is correct. This is a two-step problem. First you should set up a ratio that compares the coefficients of the balanced equation (5 O_2 and 4 CO_2) to the molar quantities that you want to know about (2.5 moles of O_2 and x moles of CO_2)

$$\frac{O_2}{CO_2} = \frac{5}{4} = \frac{2.5}{x}$$

x = 2 moles of CO_2

Now you can use the conversion formula to convert moles of CO_2 to liters of CO_2.

$$\text{Moles} = \frac{\text{liters}}{(22.4\,\text{L/mole})}$$

Liters = (moles)(22.4 L/mole)

$$\text{Liters} = \left(\frac{2\,\text{moles}}{1}\right)\left(\frac{22.4\,\text{L}}{1\,\text{mole}}\right) = 44.8 \text{ liters of } CO_2$$

20 H is correct. The limiting reactant is the reactant that disappears first in a chemical reaction. Once the limiting reactant is gone, the reaction can no longer proceed.

21 D is correct. 1 mole of $Ca(NO_3)_2$ contains 2 moles of NO_3, and 2 moles of NO_3 means 6 moles of O. Be sure to calculate the number of O atoms (6) and *not* the number of O_2 molecules (3).

22 G is correct. Because the volume doesn't change, you can ignore it and use the formula that compares pressure and temperature. Don't forget to convert degrees Celsius to Kelvin (47 + 273 = 320; 27 + 273 = 300)

$$\frac{P_1}{T_1} = \frac{P_2}{T_2}$$

$$\frac{848 \text{ mm Hg}}{320 \text{ K}} = \frac{P_2}{300 \text{ K}}$$

$P_2 = \frac{848}{320} (300) \text{ mm Hg} = 795 \text{ mm Hg}$

Even if you don't remember how to calculate the answer, you can use your intuition to rule out choice **A**. If the temperature is decreasing, then the molecules will have less energy to move around, so you know that the pressure will *decrease*, not increase. This leaves only three choices and a 1 in 3 chance.

23 **C is correct.** The Kelvin scale is the absolute temperature scale, so absolute zero is 0 K.

24 **G is correct.** According to Avogadro's law, the volume of a gas is directly proportional to the number of moles of gas present, so two samples of gas that occupy the same volume under the same conditions must contain the same number of moles of gas. None of the other choices are necessarily true.

25 **D is correct.** Of the intermolecular forces (the forces that hold liquids and solids together) listed, hydrogen bonding is the strongest, dipole-dipole interactions are next, and London dispersion forces are the weakest. If you only know the relative strengths of two out of the three forces listed, you can still use POE to eliminate some wrong answers.

26 **F is correct.** Point X is in the region of high pressure and low temperature. Those are the conditions where solids exist.

27 **C is correct.** Point Y is on the line between the gas and liquid phase. At that point, either boiling or condensation could be taking place. Boiling is the choice that is listed.

28 **F is correct.** Use the formula for specific heat capacity. $\Delta T = 22°C - 10°C = 12°C$

$q = mc\,\Delta T$

$q = (75)(4.2)(12)\text{ J} = 3780\text{ J}$

29 **B is correct.** The HCl line on the diagram is at about 70 g/100 g H_2O when the temperature is 20°C.

30 **J is correct.** At 80°C, the NH_4Cl line is the highest up on the graph of the salts listed. That means that it is the most soluble.

31 **B is correct.** The equilibrium constant places the concentrations of products in the numerator and the concentrations of reactants in the denominator. Also, the coefficients of NO_2 and N_2O_5 become exponents.

$$K_{eq} = \frac{[NO_2]^4[O_2]}{[N_2O_5]^2}$$

32 H is correct. The solubility product is the product of concentrations of particles in solution, so the larger the K_{sp}, the more particles in solution and the more soluble the salt. $Mg(OH)_2$ has the solubility product with the least negative exponent, which makes it the largest one listed. When in doubt on a question like this, you should guess either the largest or the smallest number.

33 C is correct. The equilibrium constant varies with changes in temperature but not with changes in concentration or volume.

34 F is correct. If SO_2 gas is added, it will cause crowding on the left side of the reaction, which will cause the reaction to shift to the right, creating more products. This will increase the equilibrium concentration of NO gas. All of the other choices will cause the reaction to shift towards the left. Did you use POE to eliminate wrong answers?

35 A is correct. An acidic solution has a pH of less than 7. pH is the negative logarithm of the H+ concentration, so if the pH is less than 7, 10^{-pH} will be greater than 10^{-7}. Remember, the less negative the exponent, the greater the number. If all you know is the pH of an acidic solution, you can still use POE to eliminate choices **B** and **D**.

36 J is correct. Use the formula.

$$pH + pOH = 14$$

If pH is 8, then pOH must be 6.

37 A is correct. If you draw a line straight up from the 15-mL mark on the horizontal axis, it will cross the titration curve at the point where the pH is between 2 and 3, so the pH must be 2.5.

38 J is correct. The equivalence point occurs in the middle of the steep part of the titration curve. That's the point where enough base has been added to neutralize all of the acid that was originally present. When 25 mL of KOH has been added on the curve in the question, the pH is exactly 7.

39 A is correct. Liquids are less random than gases, so a phase change from gas to liquid will produce a decrease in randomness. Randomness increases going from

$$solid \rightarrow liquid \rightarrow gas$$

40 J is correct. A decrease in temperature acts to decrease the rate of a reaction and a decrease in the concentration of reactants acts to decrease the rate of a reaction. When these two are combined, the rate of a reaction will always decrease. If you only know about temperature or concentration, but not both, you can still use POE to eliminate wrong answer choices.

41 B is correct. If ΔH is positive, the reaction is endothermic. Endothermic reactions tend to be nonspontaneous. If ΔS is negative, randomness is decreasing during the reaction, which will also tend to make the reaction nonspontaneous. Because neither ΔH and ΔS favor spontaneity, the reaction will never be spontaneous.

42 H is correct. A reaction in which two nuclei combine to create a larger nucleus is called a fusion reaction.

43 C is correct. In the experiment, only temperature is changed from trial to trial, so we can tell how temperature changes affect reaction speed. From the data, we can see that the speed of the reaction increases with increasing temperature.

44 F is correct. The data shown decrease by the same amount with each interval, so the graph is a decreasing straight line.

45 B is correct. When a metal forms an ionic bond, it gives up electrons. Electrons have negative charges, so when the metal loses a negative charge, it becomes a positive ion. Did you use POE?

46 G is correct. If the temperature of the water increases, then the dissolution of the salt must be releasing energy into the water. When energy is released in a reaction, the reaction is exothermic.

47 A is correct. When potassium (K) gives up an electron to become a positively charged ion, it has 18 electrons left. That's the same number of electrons as argon.

48 J is correct. Copper sinks in water because it has a greater density. A denser object will sink in a less dense liquid.

49 C is correct. The four elements listed all have their valence electrons in the second shell, so all of them have two electron shells, or principal energy levels. Using POE and what you know, we can start ruling out some choices. So, do atoms of different elements have the same number of electrons, number of protons, or molecular weights? Nope, so you can rule out **A**, **B**, and **D**.

50 H is correct. A catalyst increases the rate of a chemical reaction by lowering the activation energy required at the start of the reaction.

51 **A is correct.** Elements from groups 14 and 15 are nonmetals, so you can eliminate answer choice **C**, metallic. Hydrogen is found in these groups, so you can eliminate answer choice **D**, hydrogen. These elements are close together on the periodic table so they will likely form covalent bonds, answer choice **A**.

52 **J is correct.** You should be able to identify the reactants and products in the energy diagram. The products are at a lower energy than the reactants, so the reaction is exothermic.

53 **D is correct.** The table shows isotopes of the element potassium. The protons are equal, so you should choose the sample with the greatest number of neutrons. Ignore the number of electrons because they are tiny and do not affect the weight of the isotope. Sample 4 is the heaviest isotope with an atomic weight of 19 + 22 = 41.

54 **G is correct.** The graduated cylinder measures volume and the barometer measures atmospheric pressure. You can eliminate those answer choices. The triple beam balance measures the mass with much greater accuracy than the scale, so answer choice **G** is correct.

55 **C is correct.** A decrease in entropy means that the substance will become less random. The order of decreasing randomness in phases is gas to liquid to solid. Answer choice **C** is the only one that shows a decrease in entropy.

56 **H is correct.** Balance the equation to figure out the number of moles of oxygen gas. 4 moles of aluminum can form 2 moles of aluminum oxide. There are 6 moles of oxygen in aluminum oxide. Because oxygen gas is made of two moles of elemental oxygen, the coefficient for O_2 is 3.

57 **B is correct.** Use the ideal gas law to find the answer, $PV = nRT$. You are given the values for V, n, R, and T. Plug those numbers in and solve for the pressure. $P = nRT \div V = (30 \times 0.0821 \times 300) \div 1 = 738.9$.

58 **J is correct.** The thermometer is only accurate to the tenths digit. Because the thermometer is marked off to the tenths, the student is estimating the measurement in the hundredths place. This is the least accurate place, so the 5 is the least accurate digit.

59 **B is correct.** Sodium chloride has a molecular weight of 22.99 + 35.45 = 58.44 g. Divide the mass of NaCl by 58.44 to find the moles. The volume, 300 mL, is equal to 0.300 L. Divide the moles by the volume in liters to get the correct answer.

60 **F is correct.** Elements that lose two electrons to form ions are found in Group 2 on the periodic table. These are the elements beryllium, magnesium, calcium, and so on. Calcium is the only answer choice that is correct.

Practice Test 2

Periodic Table of Elements

1 H 1.0																	2 He 4.0
3 Li 6.9	4 Be 9.0											5 B 10.8	6 C 12.0	7 N 14.0	8 O 16.0	9 F 19.0	10 Ne 20.2
11 Na 23.0	12 Mg 24.3											13 Al 27.0	14 Si 28.1	15 P 31.0	16 S 32.1	17 Cl 35.5	18 Ar 39.9
19 K 39.1	20 Ca 40.1	21 Sc 45.0	22 Ti 47.9	23 V 50.9	24 Cr 52.0	25 Mn 54.9	26 Fe 55.8	27 Co 58.9	28 Ni 58.7	29 Cu 63.5	30 Zn 65.4	31 Ga 69.7	32 Ge 72.6	33 As 74.9	34 Se 79.0	35 Br 79.9	36 Kr 83.8
37 Rb 85.5	38 Sr 87.6	39 Y 88.9	40 Zr 91.2	41 Nb 92.9	42 Mo 95.9	43 Tc (98)	44 Ru 101.	45 Rh 102.	46 Pd 106.4	47 Ag 107.	48 Cd 112.	49 In 114.8	50 Sn 118.	51 Sb 121.	52 Te 127.6	53 I 126.	54 Xe 131.
55 Cs 132.	56 Ba 137.3	57 *La 138.	72 Hf 178.5	73 Ta 180.	74 W 183.9	75 Re 186.	76 Os 190.2	77 Ir 192.	78 Pt 195.1	79 Au 197.	80 Hg 200.6	81 Tl 204.4	82 Pb 207.	83 Bi 209.0	84 Po (209)	85 At (210)	86 Rn (222)
87 Fr (223)	88 Ra 226.0	89 †Ac 227.	104 Unq (261)	105 Unp (262)	106 Unh (263)	107 Uns (262)	108 Uno (265)	109 Une (267)									

*Lanthanide Series

58 Ce 140.	59 Pr 140.9	60 Nd 144.	61 Pm (145)	62 Sm 150.	63 Eu 152.	64 Gd 157.3	65 Tb 158.	66 Dy 162.	67 Ho 164.9	68 Er 167.	69 Tm 168.	70 Yb 173.0	71 Lu 175.

†Actinide Series

90 Th 232.0	91 Pa (231)	92 U 238.0	93 Np (237)	94 Pu (244)	95 Am (243)	96 Cm (247)	97 Bk (247)	98 Cf (251)	99 Es (252)	100 Fm (257)	101 Md (258)	102 No (259)	103 Lr (260)

PRACTICE TEST 2 ANSWER SHEET

Name: _____

1. Ⓐ Ⓑ Ⓒ Ⓓ
2. Ⓕ Ⓖ Ⓗ Ⓙ
3. Ⓐ Ⓑ Ⓒ Ⓓ
4. Ⓕ Ⓖ Ⓗ Ⓙ
5. Ⓐ Ⓑ Ⓒ Ⓓ
6. Ⓕ Ⓖ Ⓗ Ⓙ
7. Ⓐ Ⓑ Ⓒ Ⓓ
8. Ⓕ Ⓖ Ⓗ Ⓙ
9. Ⓐ Ⓑ Ⓒ Ⓓ
10. Ⓕ Ⓖ Ⓗ Ⓙ
11. Ⓐ Ⓑ Ⓒ Ⓓ
12. Ⓕ Ⓖ Ⓗ Ⓙ
13. Ⓐ Ⓑ Ⓒ Ⓓ
14. Ⓕ Ⓖ Ⓗ Ⓙ
15. Ⓐ Ⓑ Ⓒ Ⓓ
16. Ⓕ Ⓖ Ⓗ Ⓙ
17. Ⓐ Ⓑ Ⓒ Ⓓ
18. Ⓕ Ⓖ Ⓗ Ⓙ
19. Ⓐ Ⓑ Ⓒ Ⓓ
20. Ⓕ Ⓖ Ⓗ Ⓙ
21. Ⓐ Ⓑ Ⓒ Ⓓ
22. Ⓕ Ⓖ Ⓗ Ⓙ
23. Ⓐ Ⓑ Ⓒ Ⓓ
24. Ⓕ Ⓖ Ⓗ Ⓙ
25. Ⓐ Ⓑ Ⓒ Ⓓ
26. Ⓕ Ⓖ Ⓗ Ⓙ
27. Ⓐ Ⓑ Ⓒ Ⓓ
28. Ⓕ Ⓖ Ⓗ Ⓙ
29. Ⓐ Ⓑ Ⓒ Ⓓ
30. Ⓕ Ⓖ Ⓗ Ⓙ

31. Ⓐ Ⓑ Ⓒ Ⓓ
32. Ⓕ Ⓖ Ⓗ Ⓙ
33. Ⓐ Ⓑ Ⓒ Ⓓ
34. Ⓕ Ⓖ Ⓗ Ⓙ
35. Ⓐ Ⓑ Ⓒ Ⓓ
36. Ⓕ Ⓖ Ⓗ Ⓙ
37. Ⓐ Ⓑ Ⓒ Ⓓ
38. Ⓕ Ⓖ Ⓗ Ⓙ
39. Ⓐ Ⓑ Ⓒ Ⓓ
40. Ⓕ Ⓖ Ⓗ Ⓙ
41. Ⓐ Ⓑ Ⓒ Ⓓ
42. Ⓕ Ⓖ Ⓗ Ⓙ
43. Ⓐ Ⓑ Ⓒ Ⓓ
44. Ⓕ Ⓖ Ⓗ Ⓙ
45. Ⓐ Ⓑ Ⓒ Ⓓ
46. Ⓕ Ⓖ Ⓗ Ⓙ
47. Ⓐ Ⓑ Ⓒ Ⓓ
48. Ⓕ Ⓖ Ⓗ Ⓙ
49. Ⓐ Ⓑ Ⓒ Ⓓ
50. Ⓕ Ⓖ Ⓗ Ⓙ
51. Ⓐ Ⓑ Ⓒ Ⓓ
52. Ⓕ Ⓖ Ⓗ Ⓙ
53. Ⓐ Ⓑ Ⓒ Ⓓ
54. Ⓕ Ⓖ Ⓗ Ⓙ
55. Ⓐ Ⓑ Ⓒ Ⓓ
56. Ⓕ Ⓖ Ⓗ Ⓙ
57. Ⓐ Ⓑ Ⓒ Ⓓ
58. Ⓕ Ⓖ Ⓗ Ⓙ
59. Ⓐ Ⓑ Ⓒ Ⓓ
60. Ⓕ Ⓖ Ⓗ Ⓙ

1. Which of the following atoms contains three protons and four neutrons?

   A $^{4}_{3}\text{Li}$

   B $^{7}_{3}\text{Li}$

   C $^{10}_{7}\text{N}$

   D $^{14}_{7}\text{N}$

2. Which of the following atoms has valence electrons in the 2p subshell?

   F Li

   G Na

   H O

   J P

3. An argon atom is not very reactive mainly because—

   A it has a complete valence shell of electrons

   B it has a large number of protons

   C it has a relatively small mass number

   D its has an equal number of protons and neutrons

4. A researcher trying to identify an element found that the first two ionization energies for the element were small and of comparable size, while the third was much greater. Which of the following could be the element?

   F Ca

   G S

   H Na

   J N

5. A scientist comparing oxygen and sulfur would find that oxygen has—

   A lower electronegativity and a smaller atomic radius

   B higher electronegativity and a smaller atomic radius

   C lower electronegativity and a larger atomic radius

   D higher electronegativity and a larger atomic radius

6. A bond between an element of group 2 and an element of group 17 will be—

   F ionic

   G nonpolar covalent

   H polar covalent

   J metallic

Go On

Practice Test 2

7 What is the oxidation state of nickel in Ni_2O_3?

  A 0
  B +1
  C +2
  D +3

8 What is the name of the compound represented by CaF_2?

  F Calcium difluorine
  G Calcium fluorine(II)
  H Calcium fluorine
  J Calcium fluoride

9 What is the molecular formula of tetraphosphorous decoxide?

  A PO
  B P_4O
  C P_4O_{10}
  D PO_{10}

10 Carbonic acid forms carbon dioxide when dissolved in water. Which of the following is the formula for carbonic acid?

  F HC
  G H_2CO_3
  H HCO
  J HO

11 $\underline{}KNO_3 \rightarrow \underline{}KNO_2 + \underline{}O_2$

When the equation above is properly balanced, what is the coefficient for KNO_3?

  A 1
  B 2
  C 3
  D 4

12 Which of the following statements is true?

  F Matter can be destroyed in a combustion reaction.
  G Matter can be created in a synthesis reaction.
  H Matter is never created or destroyed in a chemical reaction.
  J Matter can be created or destroyed only in a physical change.

Go On

13 $Zn + 2\,AgCl \rightarrow ZnCl_2 + 2\,Ag$

Which of the following best describes the reaction shown above?

A Single replacement and oxidation-reduction

B Double replacement and oxidation-reduction

C Single replacement and neutralization

D Double replacement and neutralization

14 What is the value in liters of 20 milliliters?

F 0.02 L

G 0.2 L

H 2 L

J 20 L

15 A chemist measured the density of a gas using the formula below.

$$\text{Density} = \frac{\text{grams}}{\text{liters}}$$

If 40.4 grams of gas were contained in 25 liters, which of the following expresses the density of the gas with the proper number of significant figures?

A 1.616 g/L

B 1.62 g/L

C 1.6 g/L

D 2 g/L

16 What is the weight of 2 moles of carbon dioxide?

F 22.0 grams

G 44.0 grams

H 66.0 grams

J 88.0 grams

17 A solid silver bracelet weighs 22 grams. How many moles of silver are in the bracelet?

A 0.10 moles

B 0.20 moles

C 0.40 moles

D 0.80 moles

Practice Test 2

18. A researcher compared 1 mole of N_2 gas to 1 mole of H_2 gas. Which of the following statements is true if both gases were at standard temperature and pressure?

F The N_2 gas occupied a greater volume.

G The N_2 gas contained more molecules.

H The N_2 gas weighed more.

J The N_2 gas had greater average kinetic energy.

19. The coefficients of a balanced equation represent—

A the weight of the reactants and products in grams

B the volume of the reactants and products in liters

C the molar quantities of reactants and products

D the density of the reactants and products

20. $MnO_2 + 4\ HCl \rightarrow MnCl_2 + Cl_2 + 2\ H_2O$

How many grams of H_2O are produced when 1.0 mole of HCl is consumed in the reaction above?

F 4.5 grams

G 9.0 grams

H 18 grams

J 36 grams

21. An electrolytic cell was used to produce sodium. When the process was carried out, 230 grams of sodium were produced. If the theoretical yield for this process was 250 grams, what was the percent yield?

A 23%

B 46%

C 69%

D 92%

22. Before a thunderstorm, the atmospheric pressure in Richmond, Virginia, dropped to 722 mm Hg. What is this pressure in atmospheres?

F 0.72 atm

G 0.76 atm

H 0.95 atm

J 1.1 atm

23. A balloon with an initial volume of 12 L is placed under water. When this happens, the pressure increases from 760 mm Hg to 880 mm Hg. The temperature decreases from 25°C to 10°C. What is the volume of the balloon under the new conditions?

A 14.6 L

B 13.2 L

C 10.9 L

D 9.8 L

24 A tank contains three gases, O_2, H_2, and N_2. The total pressure in the tank is measured at 750 mm Hg. If the partial pressure due to O_2 is 200 mm Hg and the partial pressure due to H_2 is 420 mm Hg, what is the partial pressure due to N_2?

F 130 mm Hg
G 220 mm Hg
H 310 mm Hg
J 530 mm Hg

25 Ice floats on water because—

A ice exhibits ionic bonding and water does not
B the structure of the hydrogen bonds in ice make it less dense than water
C negative charges on the surface of ice are repelled by negative charges in water
D cold objects tend to float on warmer objects

26 Once the temperature of a solid has been raised to the melting point, the further energy required to melt it is called the—

F heat of vaporization
G heat of fusion
H activation energy
J triple point

27 Equal amounts of heat were added to equal masses of substance A and substance B. If the temperature change of substance A was *twice* as large as the temperature change of substance B, which of the following is true?

A The specific heat capacity of substance A is half as large as the specific heat capacity of substance B.
B The specific heat capacity of substance A is twice as large as the specific heat capacity of substance B.
C The initial temperature of substance A was twice as large as the initial temperature of substance B.
D The initial temperature of substance A was half as large as the initial temperature of substance B.

28 The heat of vaporization of water is 40 kJ/mole. If a 27-gram sample of water has been heated to its boiling point, how much additional heat must be added to boil the sample?

F 20 kJ
G 40 kJ
H 60 kJ
J 80 kJ

Go On

29 If 18.6 grams of Na_2O (molecular weight = 62 g/mole) are dissolved in water to form a 3.0 liter solution, what is the concentration of Na_2O?

A 0.6 M
B 0.4 M
C 0.3 M
D 0.1 M

30 What is the boiling point of a 3.0 m aqueous solution of KCl? The boiling point elevation constant for water 0.5°C/m.

F 100°C
G 103°C
H 106°C
J 112°C

31 The following reaction took place in a sealed chamber and the equilibrium concentrations were measured.

$$N_2O_4(g) \rightleftarrows 2\ NO_2(g)$$
$$[N_2O_4] = 20\ M$$
$$[NO_2] = 2\ M$$

Using the information given above, calculate the equilibrium constant for this reaction.

A $K_{eq} = 40$
B $K_{eq} = 10$
C $K_{eq} = 0.5$
D $K_{eq} = 0.2$

32 Which of the following statements is true of a reversible reaction that is at equilibrium?

F The value for the reaction quotient is greater than the equilibrium constant.
G The value for the reaction quotient is less than the equilibrium constant.
H The value for the reaction quotient is equal to the equilibrium constant.
J The reaction quotient is equal to zero.

33 The reversible reaction shown below takes place in a closed container.

$$3\ O_2(g) + heat \rightleftarrows 2\ O_3(g)$$

Which of the following changes to the equilibrium conditions would cause an increase in the concentration of O_3 gas?

A An *increase* in temperature and an *increase* in the volume of the container
B An *increase* in temperature and a *decrease* in the volume of the container
C A *decrease* in temperature and a *decrease* in the volume of the container
D A *decrease* in temperature and an *increase* in the volume of the container

Go On

34 If a solution has a pH of 4.3, what is its hydrogen ion concentration?

F $3.0 \times 10^{-3} \, M$

G $4.0 \times 10^{-4} \, M$

H $5.0 \times 10^{-5} \, M$

J $6.0 \times 10^{-6} \, M$

35 Which of the following compounds is the conjugate acid of NH_3?

A N^{3-}

B NH^{2-}

C NH_2^-

D NH_4^+

Questions 36 and 37 are based on the table below.

Acid	Formula	Acid dissociation constant, K_a
Cyanic acid	HCNO	3.5×10^{-4}
Hypobromous acid	HBrO	2.0×10^{-9}
Hydrazoic acid	HN_3	1.9×10^{-5}
Phenol acid	HC_6H_5O	1.3×10^{-10}

36 Which of the acids listed above is the *strongest* acid?

F Cyanic acid

G Hypobromous acid

H Hydrazoic acid

J Phenol acid

37 Which of the acids listed above is the *weakest electrolyte*?

A Cyanic acid

B Hypobromous acid

C Hydrazoic acid

D Phenol acid

38 A 120-mL sample of an acid solution of unknown concentration was titrated using 0.60 M NaOH solution. If the solution reached the equivalence point when 30 mL of NaOH solution was added, what was the concentration of the acid?

F 0.05 M

G 0.10 M

H 0.15 M

J 0.30 M

Questions 39 and 40 are based on the energy diagram for a reversible reaction shown below.

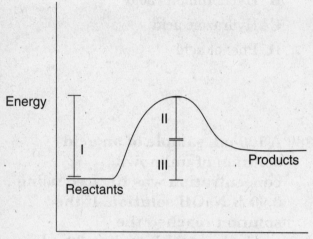

39 The forward reaction represented in the diagram is—

A an endothermic reaction
B an exothermic reaction
C both endothermic and exothermic
D neither exothermic nor endothermic

40 Which of the following will be changed by the addition of a catalyst?

F I only
G I and II only
H I and III only
J II and III only

41 Which of the following has the greatest entropy (randomness)?

A 1 mole of a solid
B 2 moles of a solid
C 1 mole of a gas
D 2 moles of a gas

42 Which of the following is the identity of particle Z in the nuclear reaction shown below?

$$^{9}_{4}Be + Z \rightarrow ^{6}_{3}Li + ^{4}_{2}He$$

F $^{1}_{1}H$

G $^{2}_{1}H$

H $^{4}_{2}He$

J $^{1}_{0}n$

43 The data below were recorded in an experiment.

Time (min.)	Pressure (mmHg)
1	20
2	40
3	80
4	160

Which of the graphs below best shows the data?

A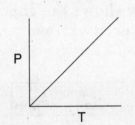

C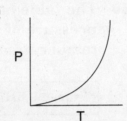

B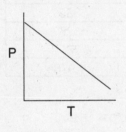

D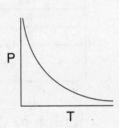

44 An experiment was conducted using the reaction shown below.

$$A + B + C \rightarrow D + E$$

Different trials were conducted in which the reactant amounts were varied as shown in the chart.

Trial	A	B	C	Temperature
1	1 mole	1 mole	1 mole	300 K
2	1 mole	2 mole	1 mole	300 K
3	1 mole	3 mole	1 mole	300 K
4	1 mole	4 mole	1 mole	300 K

This experiment will best show the effects of changes in—

F the amount of reactant A
G the amount of reactant B
H the amount of reactant C
J the temperature

45 When three liquids, A, B, and C, were poured into a beaker, they settled one on top of the other as shown in the diagram below.

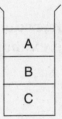

Which of the following gives the relationship of the densities of the three liquids?

A A > B > C
B C > B > A
C B > C > A
D A > C > B

Practice Test 2   243

46 A student added 36.5 grams of HCl to 2 liters of water. What is the concentration of the HCl in the solution?

F $0.5\ M$
G $1\ M$
H $2\ M$
J $4\ M$

47 An aqueous NaCl solution was compared with pure water. It was found that the NaCl solution had a—

A higher boiling point and higher freezing point
B higher boiling point and lower freezing point
C lower boiling point and higher freezing point
D lower boiling point and lower freezing point

48 An oxidation-reduction reaction involves the exchange of—

F protons
G electrons
H neutrons
J water

49 Which of the groups of elements shown below have the most similar chemical properties?

A Mg, Ca, Sr
B B, C, N
C O, F, Ne
D H, He, Li

50 The table below shows the vapor pressure of water at various temperatures.

Temperature (°C)	Vapor Pressure (mm Hg)
70	234
80	355
90	526
100	760
110	1,075

Which of the following statements is true?

F The vapor pressure of water decreases with increasing temperature.
G The vapor pressure of water is independent of temperature.
H At 100°C, water reaches its maximum vapor pressure.
J At 100°C, the vapor pressure of water is equal to the atmospheric pressure.

Go On

51 $Cl_2(g) + 2\ NO(g) \rightarrow \underline{\ ?\ }$

Chlorine gas and nitric oxide react to form what product?

A N_2OCl

B N_2O_2

C $2\ NOCl$

D $2ClO$

52 Isotopes of the same element must—

F contain equal numbers of electrons

G have the same mass numbers

H have the same number of neutrons

J contain equal numbers of protons

53 Two students are measuring a sample of calcium four times. The results are recorded in the table below.

Sample	Student 1 Mass (g)	Student 2 Mass (g)
1	16.0	16.9
2	16.2	16.2
3	16.1	17.0
4	16.1	16.8

The mass of the sample is 16.4 grams. Choose the statement that is true?

A Student 1 is accurate.

B Student 2 is precise.

C Student 2 is accurate.

D Student 1 is precise.

54 The following elements all have full octets in their valence shell *except*—

F argon

G neon

H carbon

J krypton

55

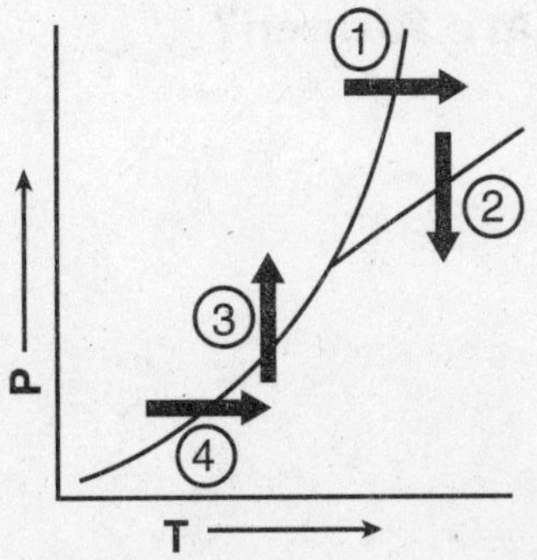

Which of the arrows above represents vaporization?

A 1

B 2

C 3

D 4

56 $\underline{\ }Fe(s) + \underline{\ }H_2O(l) \rightarrow \underline{\ }Fe_3O_4(s) + \underline{\ }H_2(g)$

The coefficients for the correctly balanced equation above are—

F 3, 4, 1, 1

G 3, 4, 1, 4

H 4, 3, 1, 3

J 4, 3, 2, 1

Practice Test 2

57 Ammonia is an important industrial chemical. The Haber process is used to make ammonia from nitrogen and hydrogen gases. The reaction of the Haber process is shown below.

$$N_2(g) + 3H_2(g) \rightleftarrows 2NH_3(g)$$

If 2,800 grams of nitrogen gas react, how many moles of ammonia are formed?

A 50
B 100
C 200
D 400

58 A scientist wishes to dissolve a substance with a heated solvent. The scientist should take all of the following precautions except—

F avoid breathing in any fumes that form
G add the solvent quickly to the substance
H swirl the solvent and substance mixture gently to combine
J hold the flask that the solvent is heated in with forceps

59 By mass, the compound NH_4 has a molecular weight of about 18 grams. What is the mass percent of each element?

A 77.8% N, 22.2% H
B 80.0% N, 20.0% H
C 22.2% N, 77.8% H
D 20.0% N, 80.0% H

60 Titration of 0.100 M HCl with 0.100 M ?

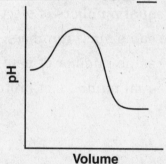

The diagram above shows the titration of hydrogen chloride with an unknown solution. Which of these could be the solution?

F NaOH
G H_2SO_4
H $HClO_3$
J CH_4

Answers and Explanations for Practice Test 2

ANSWERS AND EXPLANATIONS FOR PRACTICE TEST 2

PRACTICE TEST 2 ANSWERS

1. **B is correct.** $_3^7Li$ has three protons (atomic number) and four neutrons (mass number minus atomic number).

2. **H is correct.** Oxygen (O) has six valence electrons, two in the 2s subshell, and four in the 2p subshell.

3. **A is correct.** Atoms react and form bonds with one another in order to complete their valence shells of electrons. Argon starts with a complete valence shell, so it doesn't need to react to get one.

4. **F is correct.** There is a big jump between the second and third ionization energies, so the element probably has two valence electrons. The third electron is much more difficult to remove because it must be taken from a complete and stable shell. Ca (calcium) is the only choice listed with two valence electrons.

5. **B is correct.** Oxygen is higher up than sulfur in the same group on the periodic table. As you go down a group, atomic radius increases and electronegativity decreases, so oxygen has a smaller atomic radius and higher electronegativity.

6. **F is correct.** All of the elements of Group 2 are metals, and all of the elements of Group 17 are nonmetals. Bonds between metals and nonmetals are ionic, so the bond will be ionic.

7. **D is correct.** Oxygen takes an oxidation state of –2, and there are three oxygens in the compound. That adds up to –6. There are two nickel atoms in Ni_2O_3, so each of them must have an oxidation state of +3 to make the total of oxidation states in the compound add up to zero.

8. **J is correct.** Calcium is a metal and fluorine is a nonmetal, so CaF_2 is an ionic compound. You name an ionic compound by leaving the name of the metal alone and changing the ending of the nonmetal to -ide.

9. **C is correct.** The name of a molecular compound tells us the number of atoms of each element in the compound. So tetraphosphorous decoxide tells us that the compound contains 4 phosphorous atoms and 10 oxygen atoms, and the compound is P_4O_{10}.

10. **G is correct.** If you don't know the formula for carbonic acid, use Process of Elimination to get rid of wrong answer choices. Carbonic acid forms carbon dioxide, CO_2 when dissolved in water. You know that it has to have carbon, so you can get rid of answer choice J. Now you are left with three answer choices. Only **G**, H_2CO_3 has an acidic hydrogen.

11 **B is correct.** If there are 2 KNO_3's, then there must be 2 KNO_2's. Now you can count the O's on the both sides of the equation. There are 6 O's on the left and 4 on the right so far, so if you put 1 O_2 on the right, the equation is balanced.

$$2\ KNO_3 \rightarrow 2\ KNO_2 + 1\ O_2$$

If you chose choice **D**, you found whole number coefficients but not the *lowest* whole number coefficients. You need lowest whole number coefficients to balance an equation properly.

12 **H is correct.** It is a basic fact of chemistry that matter can never be created or destroyed in a chemical reaction or in a physical change.

13 **A is correct.** In this reaction, Zn replaces Ag in the AgCl compound, so the reaction is single replacement. Now look at the oxidation states.

$$\overset{0}{Zn} + 2\overset{1+\ \ 1-}{AgCl} \rightarrow \overset{2+\ \ 2(1-)}{Zn\ Cl2_2} + 2\overset{0}{Ag}$$

In reaction, the oxidation state of Zn goes from 0 to +2, so Zn is oxidized. The oxidation state of Ag goes from +1 to 0, so Ag is reduced. So this reaction is a single replacement reaction and an oxidation-reduction reaction.

If you know that the reaction is single replacement, but you're not sure about the rest, you can still use POE to eliminate choices **B** and **D**.

14 **F is correct.** The equation below shows how to do the conversion.

$$(20\ mL)\left(\frac{1\ L}{1,000\ mL}\right) = \left(\frac{20}{1,000}\right) L = 0.02\ L$$

15 **C is correct.** First you do the division shown in the formula.

$$Density = \frac{grams}{liters} = \frac{40.4\ grams}{24\ liters} = 1.616\ g/L$$

However, the answer can only have as many significant digits as the number in the calculation with the fewest significant digits. The number 24 has only 2 significant digits, so the answer must be rounded to 2 significant digits. When you round 1.616 to 2 significant digits, you get 1.6.

16 **J is correct.** The formula for carbon dioxide is CO_2. We can get the molecular weight of CO_2 from the periodic table.

$$12.0\ g/mole + 2(16.0\ g/mole) = 44.0\ g/mole$$

The molecular weight tells the weight of 1 mole, so the weight of 2 moles is twice as big.

$$(2 \text{ moles})(44.0 \text{ g/mole}) = 88.0 \text{ grams of } CO_2$$

17 B is correct. The chemical symbol for silver is Ag. From the periodic table, the atomic weight of Ag is 107.9. We can use the formula to convert from grams to moles.

$$\text{Moles} = \frac{\text{grams}}{\text{atomic weight}}$$

$$\text{Moles} = \left(\frac{22 \text{ grams}}{1}\right)\left(\frac{1 \text{ mole}}{107.9 \text{ grams}}\right) = 0.20 \text{ moles of Ag}$$

18 H is correct. The N_2 gas has a greater molecular weight (28 g/mole) than the H_2 gas (2 g/mole), so 1 mole of the N_2 gas will weigh more than 1 mole of the H_2 gas. Choice **F** is wrong because 1 mole of any gas will occupy 22.4 liters at STP. Choice **G** is wrong because 1 mole of anything is 6.02×10^{23} items. Choice **J** is wrong because any two gases at the same temperature will have the same average kinetic energy. Did you use POE?

19 C is correct. The coefficients of a balanced equation tell you the number of moles of reactants and products, not the mass, volume, or density.

20 G is correct. This is a two-step problem. First you should set up a ratio that compares the coefficients of the balanced equation (4 HCl and 2 H_2O) to the molar quantities that you want to know about (1 mole of HCl and x moles of H_2O)

$$\frac{HCl}{H_2O} = \frac{4}{2} = \frac{1.0}{x}$$

$x = 0.5$ moles of H_2O

Now you can use the conversion formula to convert moles of H_2O to grams of H_2O. From the periodic table, you know that the molecular weight of H_2O is 18.0 grams/mole $(2(1) + 16 = 18)$.

$$\text{Moles} = \frac{\text{grams}}{\text{atomic weight}}$$

Grams = (moles)(atomic weight)

Answers and Explanations for Practice Test 2

$$\text{Grams} = \left(\frac{0.50 \text{ moles}}{1}\right)\left(\frac{18.0 \text{ g}}{1 \text{ mole}}\right) = 9.0 \text{ grams of } H_2O$$

21 D is correct. You can use the formula for percent yield.

$$\text{Percent Yield} = \frac{\text{Actual Yield}}{\text{Theoretical Yield}} \times 100$$

$$\text{Percent Yield} = \frac{230 \text{ grams}}{250 \text{ grams}} \times 100 = 92\%$$

22 H is correct. You can convert because you know that 1 atm = 760 mm Hg.

$$722 \text{ mm Hg} \times \frac{1 \text{ atm}}{760 \text{ mm Hg}} = \left(\frac{722}{760}\right) \text{ atm} = 0.95 \text{ atm}$$

23 D is correct. Use the formula that compares pressure, volume, and temperature. Don't forget to convert from degrees Celsius to Kelvin (25 + 273 = 298, 10 + 273 = 283).

$$\frac{P_1 V_1}{T_1} = \frac{P_2 V_2}{T_2}$$

$$\frac{(760 \text{ mm Hg})(12 \text{ L})}{(298 \text{ K})} = \frac{(880 \text{ mm Hg})(V_2)}{(283 \text{ K})}$$

$$V_2 = (12 \text{ L})\left(\frac{760}{880}\right) = 9.8 \text{ L}$$

24 F is correct. According to Dalton's law, the partial pressures of all of the individual gases in a container must add up to the total pressure of the container.

$$P_{oxygen} + P_{hydrogen} + P_{nitrogen} = P_{total}$$

$$200 \text{ mm Hg} + 420 \text{ mm Hg} + P_{nitrogen} = 750 \text{ mm Hg}$$

$$P_{nitrogen} = (750 - 620) \text{ mm Hg} = 130 \text{ mm Hg}$$

25 B is correct. Water is unusual because the structure of its hydrogen bonds make the solid form less dense than the liquid form. Ice floats because it is less dense than liquid water. Did you use POE to eliminate the wrong answers?

26 **G is correct.** The heat of fusion is the heat required to break the intermolecular bonds in a solid and to convert the substance to a liquid. That's the energy required to melt the substance. Heat of vaporization is the energy required to boil a substance. Activation energy is the energy required to start a chemical reaction. The triple point is the pressure and temperature at which all three phases: gas, liquid, and solid, can exist together in equilibrium.

27 **A is correct.** The smaller the specific heat capacity, the larger the temperature change for a given amount of heat. If the temperature change of substance A was twice as big as substance B's temperature change, then substance A's specific heat capacity must be half as large as substance B's.

28 **H is correct.** First we have to convert from grams of water to moles of water.

$$\text{Moles} = \frac{\text{grams}}{\text{molecular weight}}$$

$$\text{Moles} = \left(\frac{27 \text{ grams}}{1}\right)\left(\frac{1 \text{ mole}}{18 \text{ grams}}\right) = 1.5 \text{ moles of water}$$

If it takes 40 kJ of heat to boil 1 mole of water, then it would take 60 kJ of heat to boil 1.5 moles.

29 **D is correct.** First find the number of moles of $CaBr_2$.

$$\text{Moles} = \frac{\text{grams}}{\text{molecular weight}}$$

$$\text{Moles of } Na_2O = \left(\frac{18.6 \text{ grams}}{1}\right)\left(\frac{1 \text{ mole}}{62 \text{ grams}}\right) = 0.3 \text{ moles of } Na_2O$$

Now calculate the concentration.

$$\text{Molarity} = \frac{\text{moles of solute}}{\text{liters of solution}}$$

$$\text{Molarity of } Na_2O = \frac{0.3 \text{ moles}}{3 \text{ liters}} = 0.1 \text{ M}$$

Answers and Explanations for Practice Test 2

30 G is correct. Use the formula for boiling point elevation. KCl breaks up into two particles, K⁺ and Cl⁻ so *i* is 2.

$\Delta T = ik_b m$

$\Delta T = (2)(0.5)(3.0)°C = 3°C$

The boiling point elevation is 3°C. Because the normal boiling point of water is 100°C, the new boiling point is 103°C.

31 D is correct. First set up the equilibrium expression.

$$K_{eq} = \frac{[NO_2]^2}{[N_2O_4]}$$

Now plug in the numbers.

$$K_{eq} = \frac{(2)^2}{(20)} = \frac{(4)}{(20)} = 0.2$$

32 H is correct. The reaction quotient is calculated using initial conditions. The equilibrium constant is calculated using equilibrium conditions. If the reaction is already at equilibrium, the reaction quotient will be equal to the equilibrium constant.

33 B is correct. Heat is a reactant in this reaction (the reaction is endothermic). Increasing the temperature adds heat, which causes crowding on the left side and causes the reaction to shift to the right, increasing the concentration of O_3.

There are 3 moles of gas on the right and 2 moles of gas on the left, so decreasing the volume will cause crowding among gas molecules and make the reaction shift towards the side with fewer moles of gas. That will increase the concentration of O_3.

If you understood the effects of temperature or volume, but not both, you could still use POE to eliminate some wrong answers.

34 H is correct. Use the formula. You need a calculator to get the exact value of [H⁺] here.

$[H^+] = 10^{-pH}$

$[H^+] = 10^{-4.3} = 5.0 \times 10^{-5} M$

35 D is correct. Ammonia, NH_3, is a base. When you add an H⁺ ion to a base, you get the conjugate acid. When you add an H⁺ ion to NH_3, you get NH_4^+.

36 F is correct. The strongest acid is the one that dissociates the most to create the most H⁺ ions in solution and the largest value of K_a. HCNO has the largest K_a, because its exponent is the least negative. If you weren't sure, your best bet was to guess either the largest or the smallest number.

37 D is correct. The weakest electrolyte is the substance that creates the least ions in a solution, so the weakest acid will be the weakest electrolyte. The weakest acid is the one that has the smallest value of K_a. $HC_6H_5O_2$ has the smallest K_a, because its exponent is the most negative. If you weren't sure, your best bet was to guess either the largest or the smallest number.

38 H is correct. Use the formula.

$M_A V_A = M_B V_B$

$(M_A)(120 \text{ mL}) = (0.6 \text{ }M)(30 \text{ mL})$

$M_A = 0.15 \text{ }M$

The concentration of the acid was 0.15 M.

39 A is correct. The energy of the products is greater than the energy of the reactants, so energy must have been absorbed during the reaction. When a reaction absorbs energy, it is endothermic.

40 G is correct. The activation energy for the forward reaction is represented by (I), The activation energy for the reverse reaction is represented by (II). A catalyst lowers the activation energies of both the forward and reverse reactions, so both (I) and (II) will be changed.

41 D is correct. Gases are more random than liquids or solids, and the more moles you have, the more randomness you have. Therefore, 2 moles of gas is the choice with the most randomness. Did you use POE to systematically eliminate the wrong answers?

42 F is correct. Balance the mass numbers and the charge numbers to figure out the identity of the unknown particle.

$$^{9}_{4}Be + Z \rightarrow ^{6}_{3}Li + ^{4}_{2}He$$

Mass numbers: $9 + Z_m = 6 + 4$ $Z_m = 1$

Charge numbers: $4 + Z_c = 3 + 2$ $Z_c = 1$

So, $^{1}_{1}Z$ must be $^{1}_{1}H$.

43 C is correct. The numbers shown increase by greater amounts with each interval, so the graph should be an increasing curved line.

44 **G is correct.** In the experiment, everything is kept constant from trial to trial except the amount of reactant B, so the experiment will show the effect of changes in the amount of reactant B.

45 **B is correct.** When liquids that don't mix are poured together, the densest liquid will settle to the bottom and the least dense liquid will rise to the top.

46 **F is correct.** First find out how many moles of HCl there are.

$$\text{Moles} = \frac{\text{grams}}{\text{molecular weight}}$$

$$\text{Moles} = \left(\frac{36.5 \text{ grams}}{1}\right)\left(\frac{1 \text{ mole}}{36.5 \text{ grams}}\right) = 1 \text{ moles of HCl}$$

Now figure out the concentration.

$$\text{Molarity} = \frac{\text{moles}}{\text{liters}} = \frac{1}{2} M = 0.5 \, M \text{ HCl}$$

47 **B is correct.** When a solute is added to a solvent, the solution will have a higher boiling point and a lower freezing point than the pure solvent.

48 **G is correct.** In an oxidation-reduction reaction, one compound gives electrons to another compound.

49 **A is correct.** On the periodic table, elements in vertical groups have the same number of valence electrons. Elements with the same number of valence electrons have similar chemical properties.

50 **J is correct.** At 100°C, the vapor pressure of water is 760 mm HG, which is the atmosheric pressure. When the vapor pressure of water is equal to the atmospheric pressure, the water boils.

51 **C is correct.** The number of moles of each element should balance on both sides of the chemical reaction. There are 2 moles of chlorine, 2 moles of nitrogen, and 2 moles of oxygen on the reactant side. The only answer choice with the same number of moles of each element for the product is 2 NOCl, answer choice **C**.

52 **J is correct.** All atoms of the same element have the same number of protons. Isotopes are atoms of the same element that have different numbers of neutrons. You can eliminate answer choices **G** and **H**. Isotopes do not affect the number of electrons, so the correct answer choice is **J**.

53 **D is correct.** You can see from the data table that both students measured masses that are significantly different from the actual reading, so you can eliminate any answer choice with involving accuracy. Student 1 has measurements that are close together while Student 2 has measurements that are fairly random, so Student 1 is more precise.

54 **H is correct.** Elements with full valence octets are located in Group 8, the rightmost column on the periodic table. There are also known as the noble gases because they do not tend to react with other elements. The only answer choice not in this group is carbon, answer choice **H**.

55 **B is correct.** Vaporization is the process of liquid changing to a gas or vapor. This is show as arrow 2, answer choice **B**. You may need to write the missing phases into the diagram before answering this question.

56 **G is correct.** Balance the equation by logical guessing. You should keep numbers as simple as possible by choosing 1 as the coefficient of the most complicated substance, Fe_3O_4, and then seeing if the other substances will balance. They do! You will have 3 moles of iron, 4 moles of water, and 4 moles of hydrogen gas. You can also plug in the coefficients of each answer choice to see if they will balance the equation.

57 **C is correct.** Find the number of moles of nitrogen gas that reacts: 2,800 g ÷ 28 g/mol = 100 moles. Look at the molar ratio of nitrogen gas to ammonia. It is 1 to 2, so 100 moles of nitrogen will react to form 200 moles of ammonia. Answer choice **C** is correct.

58 **G is correct.** You should never add hot liquids to another container quickly, especially if it holds another substance. This is the precaution that is incorrect. The other answer choices are safe practices in the laboratory.

59 **A is correct.** Figure out the percent mass by looking up the atomic mass of nitrogen and hydrogen. Nitrogen is about 14 gram and hydrogen is 1 gram. Then divide the atomic mass of each element by the molecular weight, 18 grams, and multiply by 100. 14 ÷ 18 × 100 = 77.8% for nitrogen. 4 ÷ 18 × 100 = 22.2% for hydrogen.

60 **F is correct.** When doing a titration, remember that you will react HCl, a strong acid, with a base. Only answer choice **F** is a base. Answer choices **G** and **H** are acids; they can donate protons. Answer choice **J** will not give off protons; it is a neutral molecule.

Answers and Explanations for Practice Test 2

Partnering with You to Measurably Improve Student Achievement

Our proven 3-step approach lets you **assess** student performance, **analyze** the results, and **act** to improve every student's mastery of skills covered by the Virginia Standards of Learning.

Assess
Deliver formative and benchmark tests

Analyze
Review in-depth performance reports and implement ongoing professional development

Act
Utilize after school programs, course materials, and enrichment resources

Order Roadmap books for your classroom or school.

1-800-REVIEW-2 • E-mail K12sales@review.com • Visit educators.princetonreview.com

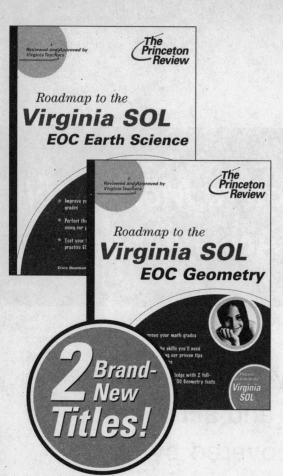

Roadmap to...
Higher Test Scores and Better Grades!

Roadmap to the Virginia SOL Series
Prepares Students for the Virginia Standard of Learning End-of-Course Tests

Roadmap to the Virginia SOL: EOC Earth Science
0-375-76441-0 • $16.00

Roadmap to the Virginia SOL: EOC Geometry
0-375-76440-2 • $16.00

Roadmap to the Virginia SOL:

EOC Algebra I
0-375-76436-4 • $16.00

EOC Algebra II
0-375-76444-5 • $16.00

EOC Biology
0-375-76443-7 • $16.00

EOC Chemistry
0-375-76442-9 • $16.00

EOC English: Reading, Literature, and Research
0-375-76439-9 • $16.00

EOC English: Writing
0-375-76445-3 • $16.00

EOC Virginia and United States History
0-375-76438-0 • $16.00

EOC World History and Geography
0-375-76437-2 • $16.00

Available at your local bookstore!